Springer
Tokyo
Berlin
Heidelberg
New York
Barcelona
Budapest
Hong Kong
London
Milan
Paris
Santa Clara
Singapore

K. Shimoji (Ed.)

Molecular Biology and Brain Ischemia

With 52 Figures Including 2 in Color

Springer

Koki Shimoji, M.D.
Professor and Chairman, Department of Anesthesiology,
Director of Intensive Care Unit and Emergency Care Unit
Niigata University School of Medicine
1-757 Asahimachi-dori, Niigata 951, Japan

ISBN-13: 978-4-431-68469-5 e-ISBN-13: 978-4-431-68467-1
DOI: 10.1007/978-4-431-68467-1

Library of Congress Cataloging-in-Publication Data

Molecular neurobiology and brain ischemia / K. Shimoji (ed.).
p. cm.
ISBN-13: 978-4-431-68469-5
1. Cerebral ischemia—Molecular aspects. 2. Molecular neurobiology. I. Shimoji, Kōki, 1935- .
[DNLM: 1. Cerebral Ischemia—physiopathology. 2. Cerebral Anoxia—physiopathology. 3. Neurobiology—methods. WL 355 M7185 1996]
RC388.5.M65 1996
616.8′1—dc20
DNLM/DLC
for Library of Congress 96-26702
CIP

Printed on acid-free paper

Typesetting: Best-set Typesetter Ltd., Hong Kong

Preface

Cerebral hypoxia/ischemia is a common cause of neurologic dysfunction following cardiopulmonary resuscitation. With the advent of aggressive resuscitation methods to treat cardiopulmonary arrest, more patients are surviving to live with posthypoxic/postischemic neurologic syndromes. Thus, there is an urgent need to understand the basic aspects of the problem. This monograph consists of reviews from scientists using advanced techniques for basic research into the molecular neurobiology of the mechanisms and effects of cerebral hypoxia/ischemia, and from clinicians actively involved in managing patients with these syndromes.

The first five chapters contain several basic topics of neurobiology to help understand brain function, followed by five chapters that deal with the mechanisms of cerebral hypoxia/ischemia and potential protective measures against it.

The first chapter describes functional proteins in the presynaptic terminal which regulate transmitter release. This is followed by a chapter demonstrating the molecular structure and physiological function of the glutamate receptor channel, which may determine the rate and quantity of calcium ion influx into the nerve cell. The third chapter surveys the metabolic glutamate responses and the intracellular mechanisms believed to play an important role in pathogenesis of hypoxic/ischemic cell injury. The fourth chapter deals with long-term potentiation and drugs, the basis for understanding memory disturbance caused by cerebral hypoxia/ischemia. The fifth chapter describes distributions of nitric oxide synthase mRNAs in the brain, currently one of the most advanced fields in neurobiology.

The sixth chapter looks at calcium-dependent neuronal cell death, which is hypothetized to be the final common effect of cerebral hypoxia/ischemia. The next chapter addresses the crucial roles of glutamate for ischemic neuronal damage. The eighth chapter describes changes in intracellular calcium ion concentration and pH brought about by ischemia. An important role of nitric oxide (NO) in ischemic brain damage is discussed in the ninth chapter. The final chapter presents evidence for the existence of an intrinsic mechanism protecting against cerebral ischemia, by loading an artificial microinjury to the brain.

It is our hope that this monograph will be worthwhile reading for those who are interested in or already engaged in the field of cerebral hypoxia/ischemia.

Koki Shimoji

Contents

List of Contributors

Functional Proteins in the Presynaptic Terminal

TERUO ABE

Summary. Neurotransmitter release from presynaptic terminals takes place rapidly by exocytosis of synaptic vesicles. It is well established that calcium ions entering presynaptic terminals through voltage-sensitive calcium channels trigger neurotransmitter release. Various lines of recent studies have led to the identification of several proteins critically involved in neurotransmitter release: SNAP-25 (synaptosome-associated protein of 25 kDa), syntaxin 1, vesicle-associated membrane protein (VAMP)/synaptobrevin, synaptotagmin, synaphin/complexin, and rab3A. These proteins form a complex that is required for docking of synaptic vesicles to, and for fusion of synaptic vesicle membranes with, the presynaptic plasma membrane. Synaptotagmin has been suggested to be a Ca^{2+} receptor in neurotransmitter release. Other proteins, including Munc-18/n-sec1/rb-sec1 and cysteine string protein, also seem to contribute to neurotransmitter release, although not present in the foregoing complex. It is very likely that neurotransmitter release is achieved by a series of rapid interactions between these proteins that is initiated by Ca^{2+}.

Key words. Neurotransmitter release—Docking/fusion complex—Syntaxin—SNAP-25—VAMP/synaptobrevin—Synaptotagmin

Introduction

It is generally assumed that higher brain functions, including learning and memory, are achieved by plasticity of synaptic transmission. Several forms of synaptic plasticity such as long-term potentiation (LTP) and long-term depression (LTD) have already been found, and their molecular mechanisms as well as their relationships to brain functions are being intensively studied. Elucidation of the mechanisms of synaptic transmission and its regulation is thus essential for understanding the higher nervous functions. Studies on the presynaptic mechanisms of synaptic transmission have been advancing much more slowly than those on the postsynaptic (neurotransmitter receptor) mechanisms, probably because of the more complex nature of presynaptic phenomena.

Department of Cellular Neurobiology, Brain Research Institute, Niigate University, 1-757 Asahimachi-dori, Niigata 951, Japan

On arrival of an action potential at the nerve terminal, neurotransmitter molecules inside synaptic vesicles are released by exocytosis. The calcium ions introduced into the nerve terminals through voltage-sensitive calcium channels trigger rapid fusion of synaptic vesicle membranes with the presynaptic plasma membrane, leading to neurotransmitter discharge. Most of the knowledge just described was acquired by electrophysiological studies (see [1]), but the molecular mechanisms underlying this process have been poorly understood. The obvious reason for this is that neurotransmitter release takes place within 1 ms through several steps involving multiple molecules. However, various studies carried out in the past 3 years have succeeded in identifying the molecules that participate in the release of neurotransmitters from nerve terminals. In this chapter, studies contributing to the identification of several proteins are briefly reviewd and the probable roles of these proteins are discussed.

Basic Properties of Neurotransmitter Release at the Nerve Terminal

Recent studies on transmitter release using capacitance measurements have presented convincing evidence for the Ca^{2+} concentration required for the release [2,3]. The concentration is rather high: maximum release at greater than 150 μM. This is in sharp contrast to the high affinity (μM or so) for Ca^{2+} in granule exocytosis in neuroendocrine cells such as in the adrenal medulla [4]. In view of the efficient Ca^{2+} removal system in nerve terminals, a low affinity for Ca^{2+} would be suitable to terminate the release promptly, as has been observed. The foregoing finding requires any candidate for the Ca^{2+} receptor in transmitter release to exert such a low affinity for Ca^{2+}. The studies mentioned have also shown that the rate of synaptic vesicle exocytosis at nerve terminals is much greater than secretory granule exocytosis in neuroendocrine cells. Thus, the synaptic vesicle exocytosis is quite distinct from granule exocytosis in neuroendocrine cells in both rate and Ca^{2+} affinity. These facts suggest unique properties of synaptic vesicle exocytosis in presynaptic terminals.

Identification of Proteins Involved in Neurotransmitter Release

The docking/fusion complex involved in neurotransmitter release has been identified through various approaches: monoclonal antibodies that immunoprecipitate N-type calcium channels, clostridial neurotoxins, intracellular membrane transport, and gene targeting and mutants. Each of these four approaches is briefly summarized next.

Monoclonal Antibodies That Immunoprecipitate Calcium Channels

By immunizing mice with synaptic plasma membranes, we obtained two monoclonal antibodies (SPM-1 and SPM-2) that immunoprecipitate ω-conotoxin-sensitive (N-type) calcium channel molecules solubilized from the brain [5]. Immunoblot analysis indicated that SPM-1 and SPM-2 recognize proteins of 36 and 28 kDa, respectively. When a digitonin extract of brain membranes was fractionated by sucrose density gradient centrifugation in the presence of digitonin, the antigens for SPM-1 and SPM-2 were mostly separate from the ω-conotoxin receptor (N-type calcium channel molecules), suggesting that neither of the antibodies recognizes subunits of the N-type calcium channel molecule but that they do recognize proteins associated with it.

On immunoaffinity chromatography of a digitonin extract of brain membranes on SPM-1 and SPM-2 Sepharose, proteins of 36, 28, and 19 kDa, respectively, were specifically retained in both cases [5]. Each of these three bands was divided into two closely spaced bands (36a, 36b; 28a, 28b; 19a, 19b) by sodium dodecyl sulfate-polyacrylamide gel electrophoresis (SDS-PAGE) in the presence of urea [6] (Fig. 1). Partial amino acid sequences of these six proteins were determined by sequencing whole proteins (for 36a and 36b) or peptides obtained by cleavage with CNBr or proteases. The two 36-kDa proteins (36a and 36b) were found to be novel proteins very similar to each other. One cDNA for the 36b protein was cloned using a mixture of oligonucleotides for a partial amino acid sequence as a probe [7]. The cDNA encoded a protein with a putative transmembrane segment at its C-terminus. Virtually identical (syntaxin 1B) and closely similar proteins (syntaxin 1A/HPC-1) were independently reported [8,9]. A recent study has shown that syntaxin 1 binds directly to the cytoplasmic portion between repeats II and III of the N-type calcium channel molecule [10].

Partial sequences of the 28a component matched exactly those of SNAP-25 (synaptosome-associated protein of 25 kDa) that had been reported previously to be localized in the presynaptic terminal [11]. Moreover, the 28a protein but not the 28b protein reacted with antisera raised against two partial sequences of SNAP-25. Thus, the 28a protein is identical to SNAP-25. The 28b component was found to be able to bind to guanosine triphosphate (GTP), and its partial sequence was exactly identical to a sequence in rab3A (smg25A) [12,13]. Moreover, the protein was specifically

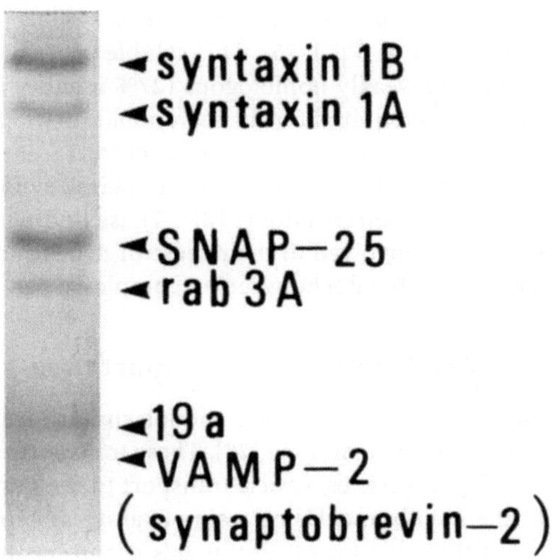

Fig. 1. Proteins specifically retained by SPM-2-Sepharose. A digitonin extract of bovine brain membranes was applied to a column of SPM-2-Sepharose, and retained materials were eluted by sodium dodecyl sulfate (SDS) (see [6])

recognized by an antiserum against rab3A, indicating that it is identical to rab3A. Despite the absence of any significantly similar amino acid sequences, both SNAP-25 and rab3A reacted with SPM-2. The 19b protein was similarly identified as the synaptic vesicle protein VAMP/synaptobrevin-2. The 19a protein differs from the 19b component (VAMP/synaptobrevin-2) because it does not react with antibodies against VAMP/synaptobrevin-2.

We have cloned a cDNA for the 19a protein using an oligonucleotide probe for a partial amino acid sequence [14]. It is a novel protein specific to the nervous system. The protein is a highly hydrophilic, acidic protein and exists predominantly in the soluble fraction of the brain. We designated the protein synaphin from the Latin *synaphia*, meaning "maintenance of the same rhythm." This protein and one closely resembling it have been independently identified by two other groups [15,16]. These proteins (termed complexins by one group) apparently bind to syntaxin [16].

Because SPM-1 does not recognize the 28-kDa or the 19-kDa proteins, and SPM-2 does not recognize the 36-kDa or the 19-kDa proteins, the retention of the six proteins by SPM-1- and SPM-2-Sepharose indicates that they exist as a complex.

A monoclonal antibody against a 58-kDa protein immunoprecipitates the N-type calcium channel molecule [17]. The protein was localized to synaptic vesicles and later found to be identical to synaptotagmin I. Synaptotagmin I binds to syntaxins 1A and 1B [8,18] and is thus indirectly associated with the N-type calcium channel. Digitonin dissociates synaptotagmin I from syntaxin 1 [8]. Synaptotagmin I coimmunoprecipitates with syntaxins 1A and 1B, SNAP-25, and VAMP/synaptobrevin-2 by SPM-1 and SPM-2 when CHAPS or Triton X-100 is used instead of digitonin [19]. Synaptotagmin I contains two C2 domains and is capable of binding Ca^{2+} and phospholipids. The affinity for Ca^{2+} is in the range of $10\,\mu M$ [20,21]. However, a very recent study [22] showed that its binding to syntaxin 1A requires a concentration of Ca^{2+} of $100\,\mu M$ or greater, matching the Ca^{2+} concentration dependence of neurotransmitter release (see earlier discussion).

A mammalian counterpart of unc-18 of *Caenorhabditis elegans* binds to syntaxins 1, 2, and 3 but not to syntaxin 4 or 5 [23–25]. This soluble protein (munc-18/n-sec1/rb-sec1) is neural specific and weakly homologous (27% identity in amino acid sequences) to yeast sec1, which is required for the secretion. Mutations of unc-18 cause abnormal accumulations of acetylcholine [26]. Furthermore, recent investigations on its mutants (rop) in *Drosophila* have shown severely impaired synaptic transmission as a result of poor release of neurotransmitters [27]. These findings strongly suggest that munc-18/n-sec1/rb-sec1 is involved in the release of neurotransmitters. Table 1 summarizes the components of the docking/fusion complex.

Studies on Intracellular Membrane Transport

Membrane proteins are transported from the endoplasmic reticulum to the Golgi apparatus and then to the plasma membrane [28]. These transports are carried out by repeating a cycle of budding of vesicles, vesicle transport to the next membrane, and fusion of vesicles with the membrane. Using a simple assay system for intracellular membrane transport that they developed, Rothman's group has been pursuing the components involved in the intracellular membrane transports. They identified NSF (*N*-ethylmaleimide-sensitive fusion protein), which is required for the membrane fusion [29]. The protein exerts a weak ATPase activity [30].

Tbale 1. Proteins in the docking/fusion complex

Protein	Molecular mass (kDa)	Localization	Cleavage site	Properties and functions
VAMP/syb-2	13	Synaptic vesicle	TeTx: Q^{76}–F^{77} BoTX/B: Q^{76}–F^{77} BoTx/D: K^{59}–L^{60} BoTx/F: Q^{58}–K^{59} BoTx/G: A^{81}–A^{82}	Binds syntaxin 1A, 1B, and SNAP-25
SNAP-25	23	Presynaptic membrane	BoTX/A: Q^{197}–R^{198} BoTx/E: R^{180}–I^{181}	Binds to syntaxin 1A, 1B, and VAMP/syb-2; attached to membrane by palmitoylation
rab3A	25	Synaptic vesicle Cytoplasm		Binds to rabphilin-3A Inhibits release?
Syntaxins 1A, 1B	33	Presynaptic membrane	BoTx/C1: K^{253}–A^{254}(1A) K^{252}–A^{253}(1B)	Binds SNAP-25, Munc-18/n-sec1, synaptotagmin, synaphin/complexion, and Ca channels
Synaptotagmin I	47	Synaptic vesicle		Binds to syntaxins 1A and 1B; a Ca sensor?
Munc-18/n-sec1/rb-sec1	68	Cytoplasm		Binds synaxins 1A and 1B. Inhibits release?
Synaphin/complexin	15	Cytoplasm		Binds syntaxins 1A and 1B
NSF	83	Cytoplasm		An ATPase; forms a trimer, binds SNAPs
α-SNAP	33	Cytoplasm		Binds NSF and SNAREs

Molecular masses are based on cDNAs. Cleavage sites represent those of rat proteins for botulinum (BoTx) and tetanus (TeTx) toxins. syb-2, Synaptobrevin-2; NSF, *N*-ethylmaleimide-sensitive fusion protein; SNAP-25, synaptosome-associated protein of 25 kDa.

NSF becomes associated with membranes in the presence of ATP and small cytoplasmic proteins called SNAPs (soluble NSF-attachment proteins) [31]. SNAPs binds to specific receptors (SNAP receptors or SNAREs) in transport vesicle membranes and other membranes. The SNAREs in the brain were searched by applying a detergent extract of brain membranes to an affinity bed consisting of NSF and SNAPs in the presence of ATP-γS [32]. Surprisingly, the SNAREs specifically retained by this affinity bed were syntaxins 1A and 1B, SNAP-25, and VAMP/synaptobrevin-2. All these proteins are localized in nerve terminals and probably participate in neurotransmitter release.

These findings strongly suggest that the mechanism underlying neurotransmitter release is very similar to those of intracellular membrane transport. Based on these results, Söllner et al. [32,33] have proposed a hypothesis for the determination of specificity of fusing membranes. This hypothesis (the SNARE hypothesis) postulates that fusion partners are determined by combination between vesicle SNAREs (v-SNAREs) and target SNAREs (t-SNAREs). Syntaxin 1 and SNAP-25 were thought to be t-SNAREs on the basis of their apparent localization in the presynaptic plasma membrane. However, recent results showed that these proteins are present not only in the presynaptic plasma membrane but also in synaptic vesicles and chromaffin granule membranes [34,35]. Therefore, the classification of these proteins as t-SNAREs may need to be revised.

Clostridial Neurotoxins

Tetanus toxins (TeTx) and seven types (A to G) of botulinum neurotoxins (BoTx) have long been known to potently block neurotransmitter release at the neuromuscular junction, but their mechanisms of action had remained a riddle. However, their primary structure deduced from cDNAs paved the way for elucidating the mechanisms. Despite their quite similar actions, the primary structures of all these toxins are not so similar to each other (maximum, 61%). However, all these toxins do contain a common structural motif, H-E-h-u-H (h, hydrophobic residue; u, uncharged residue), that is present in Zn^{2+}-dependent endopeptidase. This motif binds a Zn^{2+} and is essential for peptidase activity.

This homology led to the postulate that TeTx and BoTxs are Zn^{2+}-dependent proteases, and their target proteins in synapses were searched (for reviews, see [36] and [37]). VAMP/synaptobrevin-2 was found to be specifically cleaved by TeTx and BoTx/B, -D, and -E. Similarly, SNAP-25 and syntaxin 1 were identified as targets for BoTx/A and BoTx/C1, respectively (see Table 1). It is important to note that all the targets of clostridial toxins are contained in the protein complex described here. This fact provides convincing evidence for the pivotal role of the complex in neurotransmitter release.

All the proteins of the docking/fusion complex also exist in neuroendocrine cells such as in the adrenal medulla [38]. The existence of the docking/fusion complex itself, therofore, cannot explain the unique features of synapses in neurotransmitter release: other mechanisms are needed to explain the uniqueness.

Findings from Mutants

In addition to the munc-18/n-sec1/rb-sec1 mutants described previously, many mutants of the components of the docking/fusion complex have been obtained. Synaptic transmission in *Drosophila* mutants that lack syntaxin 1 or VAMP/synaptobrevin is abolished as a result of the block of neurotransmitter release [27,39,40]. These mutants still show docked synaptic vesicles, and spontaneous release of glutamate at the neuromuscular junction is observed in VAMP/synaptobrevin-deficient mutants but not in syntaxin 1-deficient mutants. These results suggest that syntaxin 1 and VAMP/synaptobrevin act downstream of synaptic vesicle docking [41].

In *Drosophila* and nematode (*C. elegans*) mutants that lack synaptotagmin I, synaptic transmission is seriously impaired by the poor release of neurotransmitters from nerve terminals [42–44]. However, transmitter release partially persists in these mutants. In mice whose gene for synaptotagmin I is destroyed by gene targeting, fast transmitter release in central synapses is severely suppressed while slow release is not significantly affected [45]. Docking of synaptic vesicles is apparently normal, and spontaneous release of neurotransmitters is not significantly changed.

Taken together, the results of synaptotagmin I-deficient mutant animals suggest that synaptotagmin I is indispensable for normal transmitter release. Because transmitter release partially remains in all these systems, it is possible that some other factors including other synaptotagmins are also important for release. In fact, PC12 cells lacking synaptotagmins I and II still exert normal Ca^{2+}-dependent release of catecholamines [46]. However, Ca^{2+}-dependent acetylcholine release from the same cells is abolished [47]. Acetylcholine in these cells is not stored in large dense core vesicles but in synaptic-like microvesicles. Different synaptotagmins thus may be employed for neurotransmitter release at different synapses.

How Do These Proteins Work in Neurotransmitter Release?

It is not entirely clear at present how these proteins act in the mechanism of neurotransmitter release. However, some indications for the functions of these proteins have been obtained by various studies. The following findings form the basis of the current models for neurotransmitter release.

1. Munc-18/n-secl/rb-secl is not found in the complex. However, it binds to syntaxin.
2. α-SNAP replaces synaptotagmin I from the 7 S complex consisting of syntaxin 1, SNAP-25, VAMP/synaptobrevin, and synaptotagmin.
3. The following complexes are found: syntaxin-SNAP-25-VAMP/synaptobrevin-synaptotagmin (7 S particle); syntaxin-SNAP-25-VAMP/synaptobrevin-α-SNAP; syntaxin-SNAP-25-VAMP/synaptobrevin-α-SNAP-NSF (20 S particle).

Based on these findings, initial models [24,33,48] assume an inhibitory role for synaptotagmin: synaptotagmin prevents the 7 S particle from accepting α-SNAP and NSF until the stimulus (influx of Ca^{2+}). However, very recent studies [22,45] including synaptotagmin I gene targeting strongly suggest its more active role in synaptic vesicle fusion downstream of Ca^{2+} influx. In initial models it was not clear whether NSF attachment and detachment dependent on its ATPase activity take place before or after Ca^{2+} entry. Very low ATPase activity and the very rapid nature of neurotransmitter release make it very unlikely that the low ATPase activity of NSF plays a critical role after Ca^{2+} influx. Thus, it seems reasonable to assume that NSF and SNAPs function before the entry of Ca^{2+}. NSF and SNAPs probably contribute to making synaptic vesicles and the presynaptic plasma membrane ready for fusion; then, they dissociate from the complex before Ca^{2+} influx. This state may correspond to hemifusion in which the outer leaflet of a synaptic vesicle is fused with the inner leaflet of the presynaptic plasma membrane [49].

On the basis of the Ca^{2+} dependence of its syntaxin 1 binding and severe impairment of rapid transmitter release in its gene knockout mice, synaptotagmin I is postulated to be the target for Ca^{2+} entering the nerve terminal through Ca channels [50]. The findings just described indicate that synaptotagmin I plays an important role in neurotransmitter release. Synaptotagmin I is also present in chromaffin granules in the adrenal medulla where granule exocytosis takes place in micromolar Ca^{2+}. Thus, it is very likely that different Ca^{2+} receptors act in chromaffin granule exocytosis. A less likely possibility is that a mechanism that changes the Ca^{2+} sensitivity of synaptotagmin I exists in eigher tissue.

Other Proteins Related to Neurotransmitter Release

Cysteine string protein, so called because of its cysteine-rich domain, is involved in the regulation of ω-conotoxin-sensitive calcium channels in the *Torpedo* electric organ [51] and associated with synaptic vesicles presumably by palmitoylation of several of its cysteine residues [52]. *Drosophila* mutants in the gene of this protein cause premature death and temperature-sensitive severe paralysis because of impairment of neurotransmitter release [53–55]. Cysteine string protein thus may serve as a link between synaptic vesicles and calcium channels, thereby making rapid synaptic vesicle exocytosis efficient. How the protein interacts with the docking/fusion complex is not known. VAP-33 is a newly discovered protein of 33 kDa that binds to VAMP/synaptobrevin-2 in *Aplysia* neurons. Antibodies to the protein introduced into cell bodies of *Aplysia* neurons significantly block neurotransmitter release from their presynaptic terminals [56]. The effect of this protein on the organization of the docking/fusion complex remains to be clarified.

Concluding Remarks

Several proteins that are critically involved in neurotransmitter release have been identified by recent researches based on various approaches. The remaining major problem is to elucidate the molecular cascade that leads to rapid neurotransmitter release. Undoubtedly, the regulation of neurotransmitter release via the regulation of the activities of these proteins will become a major research field for understanding the mechanisms of higher brain functions.

References

1. Katz B (1969) The release of neural transmitter substances. Liverpool University Press, Liverpool
2. von Gersdorff H, Matthews G (1994) Dynamics of synaptic vesicle fusion and membrane retrieval in synaptic terminals. Nature 367:735–739
3. Heidelberger R, Heinemann C, Neher E, Matthews G (1994) Calcium dependence of the rate of exocytosis in a synaptic terminal. Nature 371:513–515
4. Augustine G, Neher E (1992) Calcium requirements for secretion in bovine chromaffin cells. J Physiol 450:247–271
5. Saisu H, Ibaraki K, Yamaguchi T, Sekine Y, Abe T (1991) Monoclonal antibodies

immunoprecipitating ω-conotoxin-sensitive calcium channel molecules recognize two novel proteins localized in the nervous system, Biochem Biophys Res Commun 181:59–66
6. Horikawa HPM, Saisu H, Ishizuka T, Sekine Y, Tsugita A, Odani S, Abe T (1993) A complex of rab3A, SNAP-25, VAMP/synaptobrevin-2, and syntaxins in brain presynaptic terminals. FEBS Lett 330:236–240
7. Morita T, Mori H, Sakimura K, Mishina M, Sekine Y, Tsugita A, Odani S, Horikawa HPM, Saisu H, Abe T (1992) Synaptocanalin I, a protein associated with brain ω-conotoxin-sensitive calcium channels, Biomed Res 13:357–364
8. Bennett MK, Calakos N, Scheller RH (1992) Syntaxin: a synaptic protein implicated in docking of synaptic vesicles at presynaptic active zones. Science 257:255–259
9. Inoue A, Obata K, Akagawa K (1992) Cloning and sequence analysis of cDNA for a neuronal cell membrane antigen, HPC-1. J Biol Chem 267:10613–10619
10. Sheng Z-H, Rettig J, Takahashi M, Catterall WA (1994) Identification of a syntaxin-binding site on N-type calcium channels. Neuron 13:1303–1313
11. Oyler GA, Higgins GA, Hart RA, Battenberg E, Billingley M, Bloom FE, Wilson MC (1989) The identification of a novel synaptosomal-associated protein, SNAP-25, differentially expressed by neuronal subpopulations. J Cell Biol 109:3039–3052
12. Matsui Y, Kikuchi A, Kondo J, Hishida T, Teranishi Y, Takai Y (1988) Nucleotide and deduced amino acid sequences of a GTP-binding protein family with molecular weights of 25 000 from bovine brain. J Biol Chem 263:11071–11074
13. Zahraoui A, Touchet N, Chardin P, Tavitian A (1988) Complete coding sequences of the *ras*-related rab3 and 4 cDNAs. Nucleic Acids Res 16:1204
14. Ishizuka T, Saisu H, Odani S, Abe T (1995) Synaphin: a protein associated with the docking/fusion complex in presynaptic terminals. Biochem Biophys Res Commun 213:1107–1114
15. Takahashi S, Yamamoto H, Matsuda Z, Ogawa M, Yagyu K, Taniguchi T, Miyata T, Kaba H, Higuchi T, Okutani F, Fujimoto S (1995) Identification of two highly homologous presynaptic proteins distinctly localized at the dendritic and somatic synapses. FEBS Lett 368:455–460
16. McMahon HT, Missler M, Li C, Südhof TC (1995) Complexins: cytosolic proteins that regulate SNAP receptor function, Cell 83:111–119
17. Takahashi M, Arimatsu Y, Fujita S, Fujimoto Y, Kondo S, Hama T, Miyamoto E (1991) Protein kinase C and Ca^{2+}/calmodulin-dependent protein kinase II phosphorylate a novel 58-kDa protein in synaptic vesicles. Brain Res 551:279–292
18. Yoshida A, Oho C, Omori A, Kuwahara R, Ito T, Takahashi M (1992) HPC-1 is associated with synaptotagmin and ω-conotoxin receptor. J Biol Chem 267:24925–24928
19. Abe T, Saisu H, Horikawa HPM (1993) Synaptocanalins (N-type Ca channel-associated proteins) form a complex with SNAP-25 and synaptotagmin. Ann NY Acad Sci 707:373–375
20. Brose N, Petrenko AG, Südhof TC, Jahn R (1992) Synaptotagmin: a calcium sensor on the synaptic vesicle surface. Science 256:1021–1025
21. Ullrich B, Li C, Zhang JZ, McMahon H, Anderson RGW, Geppert M, Südhof TC (1994) Functional properties of multiple synaptotagmins in brain. Neuron 13:1281–1291
22. Li C, Ullrich B, Zhang JZ, Anderson RGW, Brose N, Südhof TC (1995) Ca^{2+}-dependent and -independent activities of neural and non-neural synaptotagmins. Nature 375:594–599
23. Hata Y, Slaughter CA, Südhof TC (1993) Synaptic vesicle fusion complex contains unc-18 homologue bound to syntaxin. Nature 366:347–351
24. Pevsner J, Hsu S-C, Braun JEA, Calakos N, Ting AE, Bennett MK, Scheller RH (1994) Specificity and regulation of a synaptic vesicle docking complex. Neuron 13:353–361
25. Garcia EP, Gatti E, Butler M, Burton J, DeCamilli P (1994) A rat brain secl homologue related to rop and unc18 interacts with syntaxin. Proc Natl Acad Sci USA 91:2003–2007
26. Hosono R, Hekimi S, Kamiya Y, Sassa T, Murakami S, Nishiwaki K, Miwa J, Taketo A, Kodaira K-I (1992) The unc-18 gene encodes a novel protein affecting the kinetics

of acetylcholine metabolism in the nematode *Caenorhabditis elegans*. J Neurochem 58:1517–1525
27. Harrison SD, Broadie K, van de Goor J, Rubin GM (1994) Mutations in the *Drosophila* rop gene suggest a function in general secretion and synaptic transmission. Neuron 13:555–566
28. Rothman JE (1994) Mechanisms of intracellular protein transport. Nature 372:55–63
29. Block MR, Glick BS, Wilcox CA, Wieland FT, Rothman JE (1988) Purification of an *N*-ethylmaleimide-sensitive protein catalyzing vesicular transport. Proc Natl Acad Sci USA 85:7852–7856
30. Tagaya M, Wilson DW, Brunner M, Arango N, Rothman JE (1993) Domain structure of an *N*-ethylmaleimide-sensitive fusion protein involved in vesicular transport. J Biol Chem 268:2662–2666
31. Weidman PJ, Melancon P, Block MR, Rothman JE (1989) Binding of an *N*-ethylmaleimide-sensitive fusion protein to Golgi membranes requires both a soluble protein(s) and an integral membrane receptor. J Cell Biol 108:1589–1596
32. Söllner T, Whiteheart SW, Brunner M, Erdjument-Bromage H, Geromanos S, Tempest P, Rothman JE (1993) SNAP receptors implicated in vesicle targeting and fusion. Nature 362:318–324
33. Söllner T, Rothman JE (1994) Neurotransmission: harnessing fusion machinery at the synapse. Trends Neurosci 17:344–348
34. Tagaya M, Toyonaga S, Takahashi M, Yamamoto A, Fujiwara T, Akagawa K, Moriyama Y, Mizushima S (1995) Syntaxin 1 (HPC-1) is associated with chromaffin granules. J Biol Chem 270:15930–15933
35. Walch-Solimena C, Blasi J, Edelmann L, Chapman ER, von Mullard GF, Jahn R (1995) The t-SNAREs syntaxin 1 and SNAP-25 are present on organelles that participate in synaptic vesicle recycling. J Cell Biol 128:637–645
36. Montecucco C, Schiavo G (1993) Tetanus and botulism neurotoxins: a new group of zinc proteases. Trends Biochem Sci 18:324–327
37. Niemann H, Blasi J, Jahn R (1994) Clostridial neurotoxins: new tools for dissecting exocytosis. Trends Cell Biol 4:179–185
38. Burgoyne RD, Morgan A (1995) Ca^{2+} and secretory-vesicle dynamics. Trends Neurosci 18:191–196
39. Schulze KL, Broadie K, Perin MS, Bellen HJ (1995) Genetic and electrophysiological studies of *Drosophila* syntaxin-1A demonstrate its role in nonneuronal secretion and its essential role in neurotransmitter release. Cell 80:311–320
40. Sweeney ST, Broadie K, Keane J, Niemann H, O'Kane CJ (1995) Targeted expression of tetanus toxin light chain in *Drosophila* specifically eliminates synaptic transmission and causes behavioral defects. Neuron 14:341–351
41. Broadie K, Prokop A, Bellen HJ, O'Kane CJ, Schulze KL, Sweeney ST (1995) Syntaxin and synaptobrevin function downstream of vesicle docking in *Drosophila*. Neuron 15:663–667
42. DiAntonio A, Schwarz TL (1994) The effect on synaptic physiology of synaptotagmin mutations in *Drosophila*. Neuron 12:909–918
43. Littleton JT, Stern M, Perin MS, Bellen HJ (1994) Calcium dependence of neurotransmitter release and rate of spontaneous vesicle fusions are altered in *Drosophila* synaptotagmin mutants. Proc Natl Acad Sci USA 91:10888–10892
44. Nonet ML, Grundahl K, Meyer BJ, Rand JB (1994) Synaptic function is impaired but not eliminated in *C. elegans* mutants lacking synaptotagmin. Cell 73:1291–1305
45. Geppert M, Goda Y, Hamme RE, Li C, Rosahl TW, Stevens CF, Südhof TC (1994) Synaptotagmin I: a major Ca^{2+} sensor for transmitter release at a central synapse. Cell 79:717–727
46. Shoji-Kasai Y, Yoshida A, Sato K, Hoshino T, Ogura A, Kondo S, Fujimoto Y, Kuwahara R, Kato R, Takahashi M (1992) Neurotransmitter release from synaptotagmin-deficient clonal variants of PC12 cells. Science 256:1820–1823
47. Bauerfeind R, Jelinek R, Huttner WB (1995) Synaptotagmin I- and II-deficient PC12 cells exhibit calcium-independent, depolarization-induced neurotransmitter release

from synaptic-like microvesicles. FEBS Lett 364:328–334
48. Söllner T, Bennett MK, Whiteheart SW, Scheller RH, Rothman JE (1993) A protein assembly-disassembly pathway in vitro that may correspond to sequential steps of synaptic vesicle docking activation, and fusion. Cell 75:409–418
49. Monck JR, Fernandes JM (1994) The exocytotic fusion pore and neurotransmitter release. Neuron 12:707–716
50. Südhof TC (1995) The synaptic vesicle cycle: a cascade of protein–protein interactions. Nature 375:645–653
51. Gundersen CB, Umbach JA (1992) Suppression cloning of the DNA encoding a candidate presynaptic calcium channel subunit of *Torpedo*. Neuron 9:527–537
52. Mastrogiacomo A, Parson SM, Zampighi GA, Jenden DJ, Umbach JA, Gundersen CB (1994) Cysteine string proteins: a potential link between synaptic vesicles and presynaptic Ca^{2+} channels. Science 263:981–982
53. Umbach JA, Zinsmaier KE, Eberle KK, Benzer S, Gundersen CB (1994) Presynaptic dysfunction in *Drosophila* csp mutants. Neuron 13:899–908
54. Zinsmaier KE, Eberle KK, Buchner E, Walter N, Benzer S (1994) Paralysis and early death in cysteine string protein mutants of *Drosophila*. Science 263:977–980
55. Zinsmaier K, Hofbauer A, Heimbeck G, Pflugfelder GO, Buchner S, Buchner E (1990) A cysteine string protein is expressed in retina and brain. J Neurogenet 7:15–29
56. Skehel PA, Martin KC, Kandel ER, Bartsch D (1995) A VAMP-binding protein from *Aplysia* required for neurotransmitter release. Science 269:1580–1583

Molecular Structure and Physiological Function of the Glutamate Receptor Channel

KENJI SAKIMURA

Summary. We have identified 16 kinds of mouse glutamate receptor channel subunits by molecular cloning. These subunits can be classified into six subfamilies according to their amino acid sequence homology. Functional analysis of those subunits showed that this grouping corresponds very well to the pharmacological classification. We have demonstrated that *N*-methyl-D-aspartate (NMDA) receptor channels with high activity are formed by coexpression of the ε- and ζ-subunit families. The four ε-subunits are distinct in distribution, functional properties, and regulation. The functional properties of the NMDA receptor channel are critically determined by the constituting ε-subunits. To examine the physiological roles in vivo of the diverse ε-subunits of the NMDA receptor channel, we employed a gene targeting technique. Disruption of the ε1-subunit did not appreciably affect the growth and mating of the mice, but resulted in significant reduction of the NMDA receptor channel current and long-term potentiation at the hippocampal CA1 synapses. The ε1 mutant mice also showed a moderate deficiency in the ability of learning that was estimated by the Morris water-maze task. These results support the concept that NMDA receptor channel-dependent synaptic plasticity is the cellular basis of certain forms of learning.

Key words. Glutamate receptor channel—NMDA receptor channel—Cloning—Heteroligomer—LTP—Plasticity—Learning—Memory—Gene targeting

Introduction

Glutamate receptors are found throughout the mammalian brain, where they constitute the major excitatory transmitter system. The well-known and best-studied glutamate receptors are called ligand-gated ion channels or ionotropic glutamate receptor, which are permeable to cations. They have traditionally been classified, by electrophysiological and pharmacological studies, into three broad subtypes: the α-amino-3-hydroxy-5-methyl-4-isoxazole propionic acid (AMPA), kainate, and *N*-methyl-D-aspartate (NMDA) receptors [1,2]. Recently, a family of G-protein-

---

Department of Cellular Neurobiology, Brain Research Institute, Niigata University, 1-757 Asahimachi-dori, Niigata 951, Japan coupled glutamate receptors, which are also called metabotropic glutamate or *trans*-1-aminocyclopentane-1,3-dicarboxylate (TACPD) receptors, was identified [3].

The glutamate receptor (GluR) channel plays a key role in brain function. Most of the fast excitatory synaptic transmission is mediated by GluR channels in the central nervous system [1]. GluR channels are also essential for activity-dependent synaptic plasticity such as long-term potentiation (LTP) and long-term depression (LTD), which are thought to underlie memory acquisition and learning [4,5]. Furthermore, cumulative evidence suggests that GluR channels are involved in experience-dependent synaptic plasticity in the developing brain [6] and that they play a role in many brain diseases and the damage that occurs after head injuries. It has also been known that glutamate is toxic to neurons in culture and in vivo.

Many experiments have suggested that the glutamate receptor is a mediator of the toxic effects [7,8]. These observations have led investigators to think that many neurological accidents involving strokes and epileptic seizures may result in brain damage because of overstimulation by glutamate. It has also been proposed that degenerative diseases such as Alzheimer's disease, Huntington's disease, and Parkinson's disease may involve neuronal cell death caused by excessive activation of the glutamate receptor system [8]. Thus, extensive study of the glutamate receptor systems is crucial for the understanding of basic brain functions such as learning and memory as well as for rational treatment of brain diseases.

Molecular Diversity of the GluR Channel

The first GluR channel subunit, designated GluR1, was isolated from rat brain by Hollmann et al. using an elegant expression cloning strategy [9]. We have identified 16 species of mouse GluR channel subunits by successive screening of mouse brain cDNA libraries under low stringency conditions with an α1-subunit (the mouse counterpart of the GluR1 subunit) cDNA [10] and newly identified subunit cDNAs as probes [11–18]. The entire primary structures of these subunits were deduced by nucleotide sequence analysis of the cloned cDNAs. GluR channel subunits have a putative amino-terminal signal peptide and four hydrophobic segments (M1–M4) in the middle of the molecules. These structural characteristics seem to be common to neurotransmitter-gated ion channels, such as the nicotinic acetylcholine receptor channel [19], the γ-aminobutyric acid (GABA) receptor channel [20], and the glycine receptor channel [21]. The four hydrophobic segments are well conserved among subunits, and segment M2 is assumed to constitute the inner wall of the ion channel in analogy of the acetylcholine receptor channel. Recently, however, a three-transmembrane segment model was proposed in which a putative channel-lining segment, M2, loops into the membrane without traversing it [22].

We have classified these subunits into six subfamilies according to the amino acid sequence identity (Table 1); the rat counterparts are differently named by the groups of Heinemann, Seeburg, and Nakanishi [23]. The amino acid sequence identity is as high as ~40% to ~70% within a subfamily, but is as low as ~10% to ~40% between subfamilies. The presence of variants produced most likely by alternative splicing is noted in some subunits [23].

Table 1. Subunits of the rodent glutamate receptor (GluR) channel

Subfamily	Subunit	Function	Rat counterpart
GluRα	α1, α2, α3, α4	AMPA/Kainate	GluR1–GluR4 (Heinemann) GluRA–GluRD (Seeburg)
GluRβ	β1, β2, β3	Kainate	GluR5–GluR7 (Heinemann)
GluRγ	γ1, γ2	Kainate	KA1, KA2 (Seeburg)
GluRδ	δ1, δ2	Unknown	δ1, δ2 (Seeburg)
GluRε	ε1, ε2, ε3, ε4	NMDA	NR2A–NR2D (Seeburg)
GluRζ	ζ1	NMDA	NMDAR1 (Nakanishi)

AMPA, α-Amino-3-hydroxy-5-methyl-4-isoxazole proprionic acid; NMDA, *N*-methyl-D-aspartate.

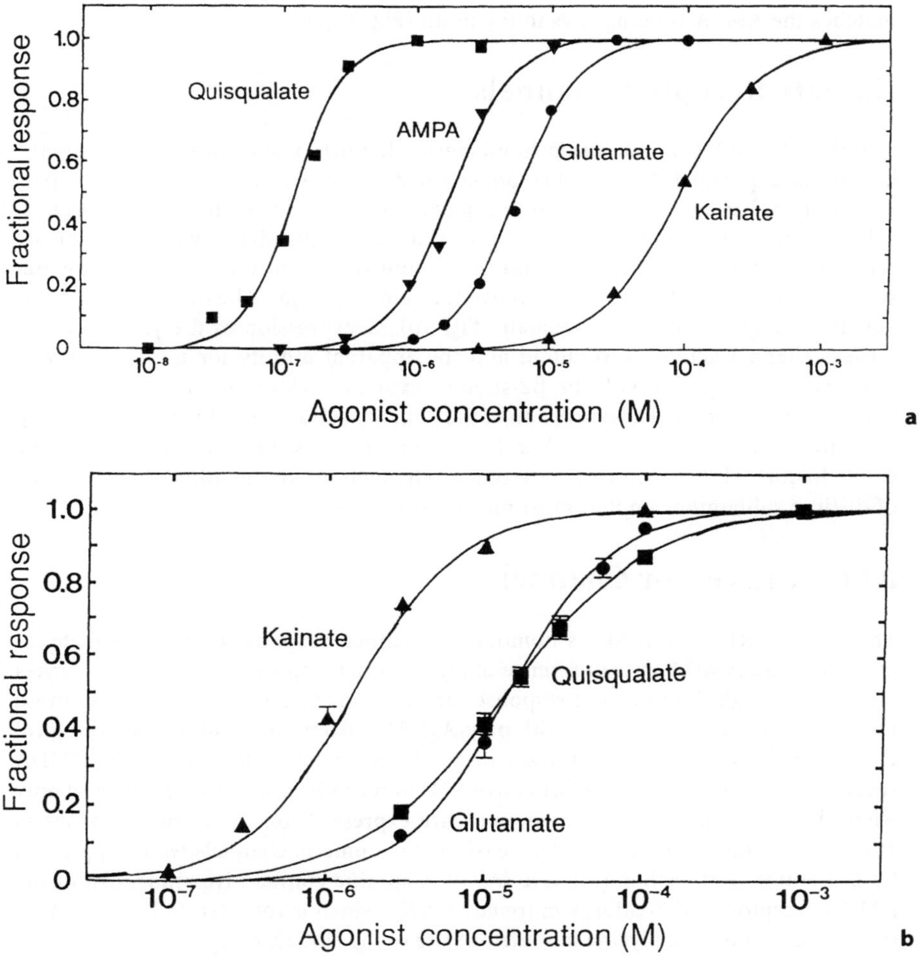

Fig. 1a,b. Dose–response curves for agonists of the α1/α2 (**a**) and β2/γ2 (**b**) heteromeric channels expressed in *Xenopus* oocytes. AMPA, α-Amino-3-hydroxy-5-methyl-4-isoxazole proprionic acid; NMDA, *N*-methyl-D-aspartate. (Modified from [10] and [12])

AMPA/Kainate Receptor Channels

Functional properties of the respective subunits were examined after expression in *Xenopus* oocytes by injection of the subunit-specific mRNAs synthesized in vitro using cloned cDNAs as templates. The α1-subunit alone forms functional GluR channels, but the current responses are much stronger when the α1- and α2-subunits are expressed together [10]. The increases in channel activity and Hill coefficient values for the dose–response curves for agonists indicate a positive cooperative interaction between the α1- and α2-subunits. The apparent affinities for agonists of the α1/α2 heteromeric channel as well as the α1 homomeric channel are quisqualate > AMPA > L-glutamate > kainate (Fig. 1a). No measurable response is detected for NMDA plus glycine. Thus, the subfamily represents the AMPA-selective GluR channel. The distribution of the four subunit mRNAs revealed by in situ hybridization largely matches the AMPA-binding sites in the brain (Fig. 2) [24].

Kainate Receptor Channels

The β2- (GluR6) subunit forms homomeric channels responsive to L-glutamate, kainate, and quisqualate, but not responsive to AMPA [11,12,25]. The order of apparent affinity is kainate > quisqualate > L-glutamate. On the other hand, the γ2-(KA-2) subunit does not form functional receptor channels responsive to various glutamate receptor agonists. However, when the γ2-subunit is expressed together with the β2-subunit, it forms GluR channels selective for kainate [12,26]. The order of potency is kainate > L-glutamate = quisqualate (Fig. 1b). Coexpression of the γ2-subunit increases channel activity severalfold and the apparent affinity for L-glutamate and quisqualate, compared with the β2-subunit expressed alone. In situ hybridization studies suggest regional coexpression of the β- and γ-subunit mRNAs (Fig. 2) [25–28]. Therefore, the members of the β and γ subfamilies constitute the kainate-selective GluR channel. The formation of heteromeric channels by combination of the subunits of distinct subfamilies implies great molecular diversity in GluR channels.

NMDA Receptor Channel

The ζ1- or NR1-subunit forms homomeric channels responsive to L-glutamate, L-aspartate, and NMDA in the presence of glycine when expressed in *Xenopus* oocytes [15,29], although their current responses are very small, even compared with those observed in the oocyte injected with poly(A)$^+$ RNA from the brain. Pharmacological characteristics of the ζ1-subunit are in good agreement with those of the NMDA receptor channel. However, highly active NMDA receptor channels are observed only when distantly related ζ- and ε-subunits are expressed together. The responses to 10 μ*M* L-glutamate plus 10 μ*M* glycine of the heteromeric channels were suppressed by D-2-amino-5-phosphonovalerate (APV), a specific competitive antagonist of the NMDA receptor, and 7-chlorokynurenate (7CK), which is reported to be a competitive antagonist for the glycine modulatory site of the NMDA receptor. The responses were inhibited also by noncompetitive antagonists such as Mg^{2+}, Zn^{2+}, and (+)-MK-801. These channels exhibited clear inward currents in Na^+- and K^+-free Ringer's solution containing 20 m*M* Ca^{2+}, whereas a marginal outward current was observed in control Na^+- and K^+-free Ringer's solution, indicating that the heteromeric channels

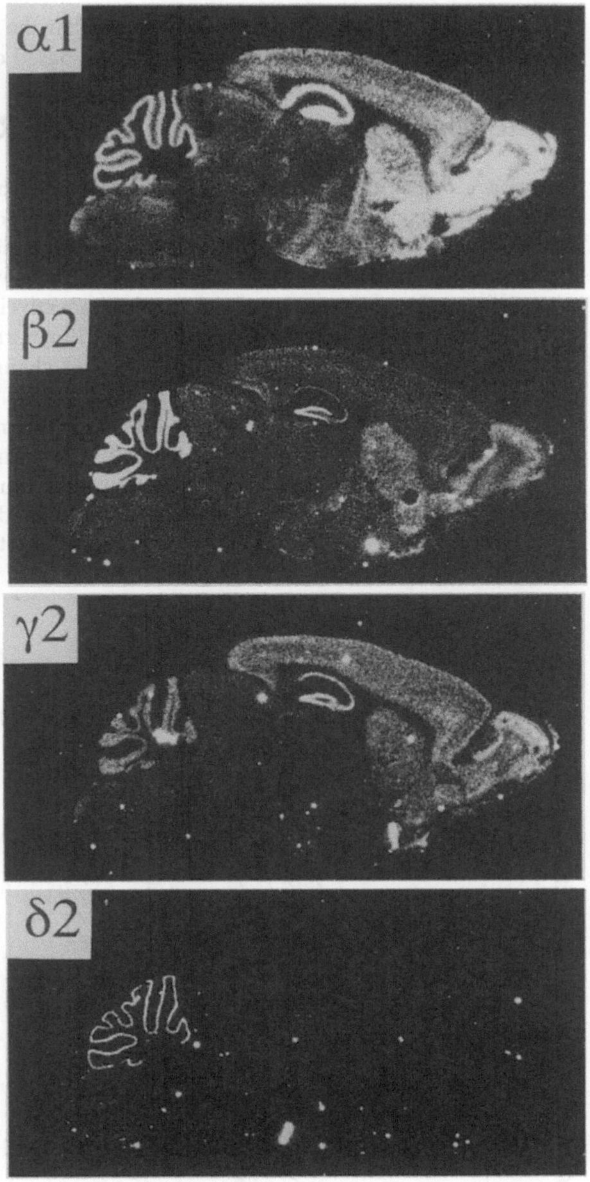

FIG. 2. Expression of α1-, β2-, γ2-, and δ2-subunit mRNAs in the developing mouse brain. Images of in situ hybridization of dark-field micrographs were taken from emulsion-dipped parasagittal sections

are permeable to Ca^{2+}. These results suggest that the nature of the NMDA receptor channel is heteromeric and that its diversity is based on the molecular level.

Functional Diversity of the NMDA Receptor Channel

Pharmacological properties of the ε1/ζ1, ε2/ζ1, ε3/ζ1, and ε4/ζ1 channels were examined by analyzing the dose–response relationships for L-glutamate and glycine in Ba^{2+}-Ringer's solution to minimize the effect of secondarily activated Ca^{2+}-dependent Cl^- currents (Fig. 3). The EC_{50} values for L-glutamate were 1.7, 0.8, 0.7, and 0.4 μ*M* for the ε1/ζ1, ε2/ζ1, ε3/ζ1, and ε4/ζ1 channels, respectively, while those for glycine were 2.1, 0.3, 0.2, and 0.09 μ*M*, respectively [17,18]. Furthermore, the sensitivity to APV is of the order of the ε1/ζ1 > ε2/ζ1 > ε3/ζ1 > ε4/ζ1 channels, while that to 7CK is of the order of the ε3/ζ1 > ε2/ζ1 > ε1/ζ1 = ε4/ζ1 channels. The ε1/ζ1 and ε2/ζ1 channels are more sensitive to Mg^{2+}, Zn^{2+}, and (+)-MK-801 than the ε3/ζ1 and ε4/ζ1 channels, but the sensitivities to phencyclidine (PCP), ketamine, and *N*-allylnormetazocine (SKF-10,047) are only slightly variable among the four ε/ζ channels [17,18,30]. The activity of the NMDA receptor channel is modulated by protein kinases and protein phosphatases. Treatment of oocytes with 12-*o*-tetradecanoylphorbol-13-acetate (TPA) potentiated the responses of the ε1/ζ1 and ε2/ζ1 channels, but did not augment

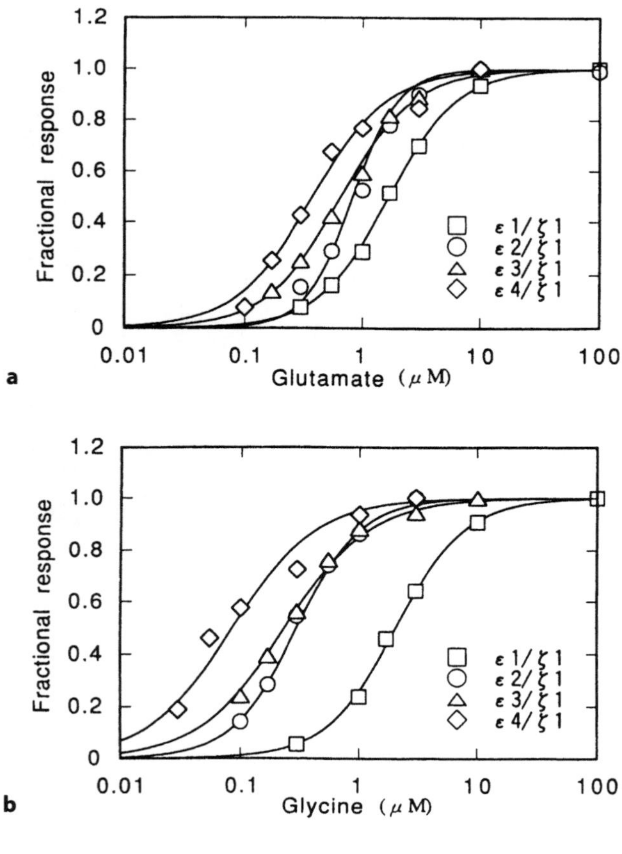

Fig. 3a,b. Dose–response curves for agonists of the heteromeric NMDA receptor channel. Dose–response curves for L-glutamate of the ε1/ζ1 (*squares*), ε2/ζ1 (*circles*), ε3/ζ1 (*triangles*), and ε4/ζ1 (*diamonds*) channels were obtained in the presence of 10 μ*M* glycine (**a**); those for glycine were obtained in the presence of 10 μ*M* L-glutamate (**b**). (Modified from [17] and [18])

the responses of the ε3/ζ1 and ε4/ζ1 channels [17,31]. These results suggest that pharmacologically distinct NMDA receptor channels can be formed depending on the constituting ε-subunits.

Distribution of the NMDA receptor channel subunit mRNAs in mouse brain was examined by in situ hybridization analyses (Fig. 4) [32,33]. The ε1-subunit mRNA is expressed postnatally and widely in the brain; in particular, the level of expression is higher in the cerebral cortex, the hippocampal formation, and cerebellar granule cells. On the other hand, ε2-subunit mRNA is found throughout the entire embryonic brain, but its expression becomes restricted to the forebrain at postnatal stages. High levels of the ε2-subunit mRNA expression are observed in the cerebral cortex, hippocampal formation, septum, caudate-putamen, olfactory bulb, and thalamus. The ε3-subunit mRNA appears postnatally and predominantly in the cerebellum. It is strongly expressed in the granule cell layer of the cerebellum but is weakly detected in the olfactory bulb and the thalamus. The ε4-subunit mRNA is abundantly expressed in the diencephalon and the brainstem at embryonic and neonatal stages, but its expression decreases to low levels before postnatal day 14. By contrast, the ζ1-subunit mRNA is distributed ubiquitously in the brain throughout development.

These findings suggest that changes in the subunit composition of the NMDA receptor channel take place during brain development. Furthermore, alteration of the subunit composition would produce changes in the functional properties of the NMDA receptor channel. In accord with this idea, it has been reported that the pharmacological and electrophysiological properties of the NMDA receptor, such as the sensitivity to Mg^{2+} block, the apparent affinity for glycine, and the channel kinetics, undergo changes during brain development [34–36].

A Single Amino Acid Determines the Channel Properties of the GluR Channel

The α1/α2 heteromeric channel exhibits a large inward current at −70-mV membrane potential in normal Ringer's solution, but it shows no inward current in Na^+- and K^+-free Ringer's solution containing 20 m*M* Ca^{2+}. Thus, the heteromeric channel is essentially impermeable to Ca^{2+}, like most native AMPA receptor channels. On the other hand, the α1 homomeric channel exhibits clear inward currents in both normal and Ca^{2+}-Ringer's solutions. These observations indicate that the α2-subunit regulates the ion selectivity of the AMPA-selective GluR channel. Inspection of the amino acid sequence of the α2-subunit reveals the unique presence of arginine 586 in the putative channel-forming segment M2. To examine whether this positively charged residue plays a role in selective cation permeation through the channel, we introduced a point mutation into the α2-subunit to substitute glutamine for arginine (mutation α2-R586Q). The heteromeric channel, composed of the wild-type α1-subunit and the mutant α2-R586Q subunit, is highly permeable to Ca^{2+}. These results show that arginine 586 in segment M2 of the α2-subunit is a critical determinant of the Ca^{2+} permeability of the AMPA-selective GluR channel [37].

The voltage-dependent Mg^{2+} block is the key to the depolarization-dependent activation of the NMDA receptor channel, which is the basis of an activity-dependent change of synaptic efficacy. All subunits of the NMDA receptor channel possess asparagine in segment M2 at the position corresponding to glutamine or arginine that determines the Ca^{2+} permeability of the AMPA-selective GluR channel. The wild-type

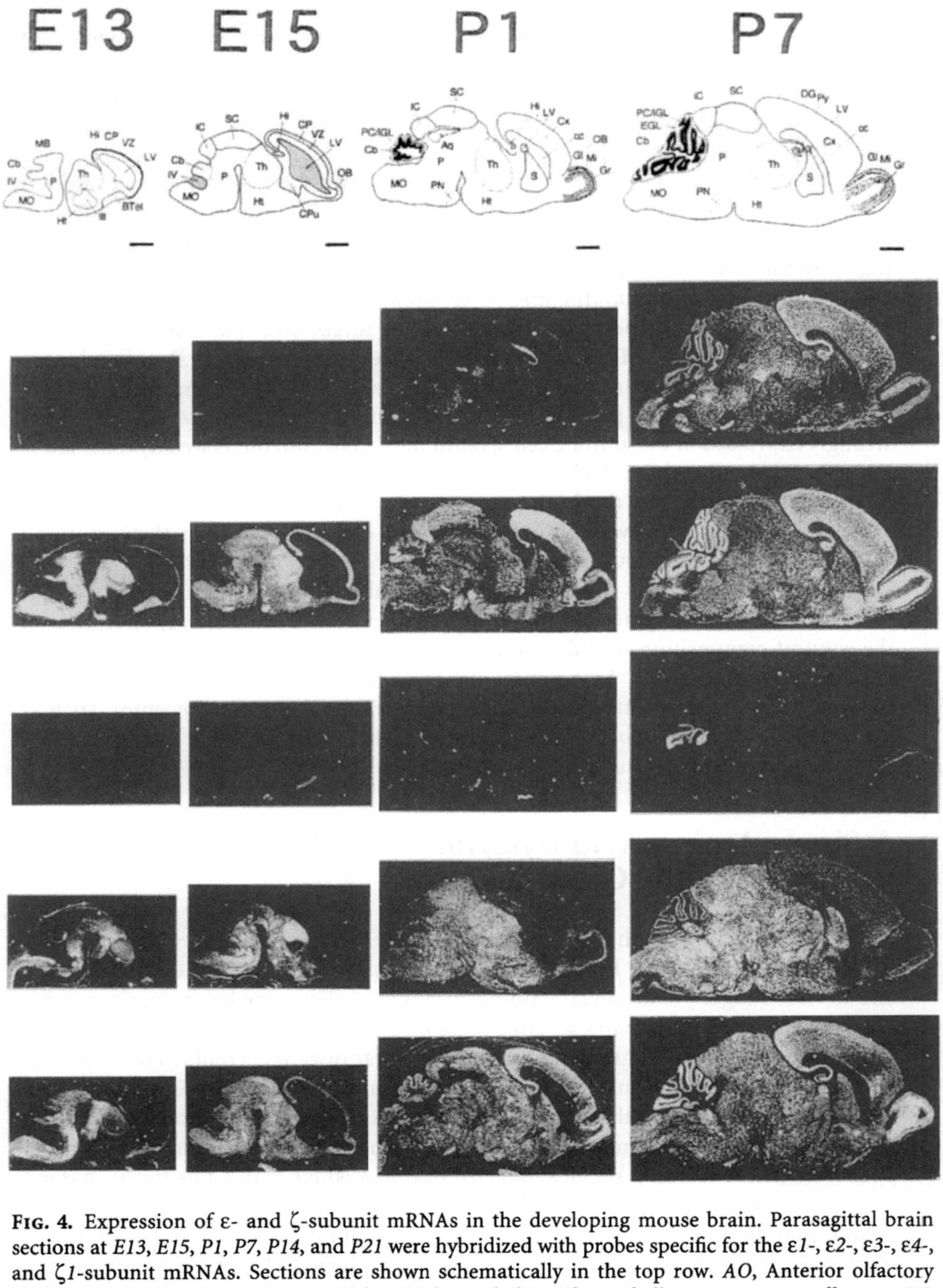

a

FIG. 4. Expression of ε- and ζ-subunit mRNAs in the developing mouse brain. Parasagittal brain sections at *E13, E15, P1, P7, P14,* and *P21* were hybridized with probes specific for the ε*1*-, ε*2*-, ε*3*-, ε*4*-, and ζ*1*-subunit mRNAs. Sections are shown schematically in the top row. *AO*, Anterior olfactory nucleus; *Aq*, cerebral aqueduct; *BTel*, basal telencephalon; *Cb*, cerebellum; *cc*, corpus callosum; *CN*, cerebellar nuclei; *cp*, cortical plate; *CPu*, caudate-putamen; *Cx*, cerebral cortex; *DG*, dentate gyrus; *EGL*, external granular layer; *GI*, olfactory glomerular layer; *Gr*, olfactory granular layer; *Hi*, hippocampus; *Ht*, hypothalamus; *IC*, inferior colliculus; *IZ*, intermediate zone; *LV*, lateral ventricle; *MB*, midbrain; *Mi*, olfactory mitral cell layer; Mol, cerebellar molecular layer; *MO*, medulla oblongata; *OB*, olfactory bulb; *P*, pons; *PC/Gr*, Purkinji cell layer/granule cell layer; *PC/IGL*, Purkinje cell layer/internal granular layer; *PN*, pontine nuclei; *Py*, hippocampal pyramidal cell layer; *S*, septum; *SC*, superior colliculus; *Th*, thalamus; Tu, olfactory tubercle; *VZ*, ventricular zone; *III*, third ventricle; *IV*, fourth ventricle. (From [33], with permission)

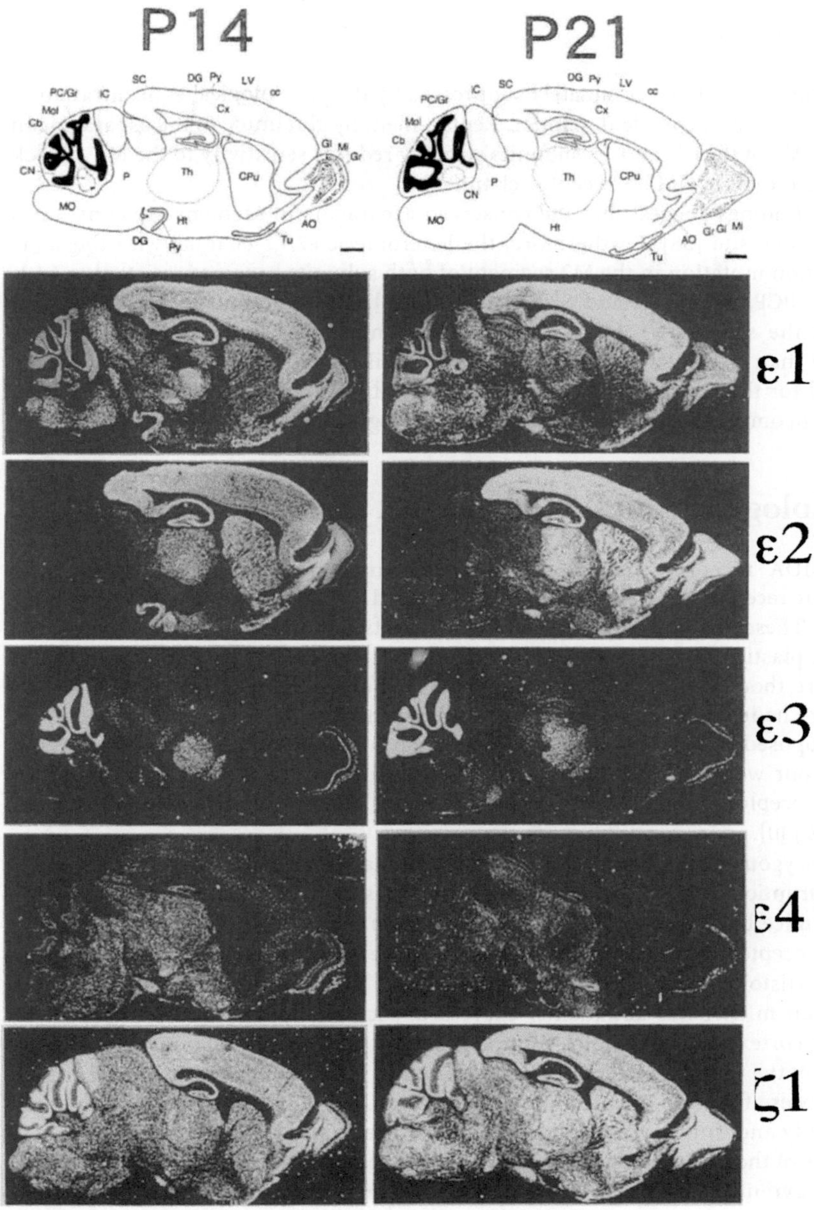

Fig. 4. *Continued*

ε2/ζ1 channel is almost completely suppressed by the physiological concentration of Mg^{2+} in a voltage-dependent manner. Replacement by glutamine of the asparagine in segment M2 of the ε2- and ζ1-subunits strongly reduces sensitivity to the Mg^{2+} block of the heteromeric NMDA receptor channel.

These findings suggest that the conserved asparagine in segment M2 constitutes a Mg^{2+} block site [38]. Furthermore, the heteromeric ε2/ζ1 channel with the same substitution mutation in the M2 segment of both subunits becomes resistant to (+)-MK-801, PCP, ketamine, and SKF-10,047 [30,38]. The mutation in the ζ1-subunit reduced the sensitivity of the heteromeric channels to ketamine and PCP more strongly than that in the ε2-subunit [30,38], while mutations in both subunits are required for the (+)-MK-801 resistance. These observations suggest that the site of these noncompetitive antagonists overlaps the Mg^{2+} block site.

Physiological Roles of the NMDA Receptor Channel

The NMDA receptor channel is unique in functional properties among other glutamate receptors. It is gated both by ligands and by voltage and is highly permeable to Ca^{2+}. These characteristics directly relate to its important physiological roles in synaptic plasticity as a molecular coincidence detector. Some forms of LTP and LTD, which are thought to underlie learning and memory, are dependent on the NMDA receptor channel. On the basis of the Mg^{2+} sensitivity and expression patterns, we have proposed that the ε1- and ε2-subunits play a role in synaptic plasticity [17,39]. To test our working hypothesis and to investigate the functional roles in vivo of NMDA receptor channel diversity, we generated mutant mice defective in the ε1-subunits [40].

Homozygous GluRε1 mutant mice grew and mated normally, while the ζ1 (NR1) gene-disruption mice died soon after birth. In situ hybridization analysis of the mutant mice brain showed that expressions of mRNAs corresponding to the other NMDA receptor channel subunits were not appreciably affected by the GluRε1 disruption. Histological examination was carried out in various brain regions where the ε1-subunit mRNA was abundantly expressed, such as the hippocampal formation, cerebral cortex, olfactory bulb, ventral posterior nucleus of thalamus, spinal tract nucleus of the trigeminal nerve, and inferior olive. No obvious morphological abnormalities were found in the ε1 mutant mice. The ε1-subunit mRNA is expressed only postnatally and thus may exert little effect on development [32].

Effects of the ε1 deprivation on excitatory synaptic transmission in the hippocampal CA1 pyramidal neurons were examined. There were no significant differences in either the transmission efficacy or the extent of paired-pulse facilitation between the wild-type and mutant slices. However, NMDA receptor channel currents, expressed as the ratio to 6-cyano-7-nitroquinoxaline-2,3-dione- (CNQX-) sensitive non-NMDA receptor channel currents, were significantly reduced in the mutant slices compared with the wild-type slices. The residual NMDA receptor channel activity probably can be ascribed to the presence of the ε2-and ζ1-subunits. We then examined synaptic plasticity in the hippocampal CA1 region by field potential recordings. Figure 5 shows that tetanic stimulation evoked LTP both in the wild-type and in the mutant slices, but the extent of the potentiation was significantly decreased in the mutant slices. Because it is well established that Ca^{2+} entry through NMDA receptor channels triggers induction of LTP in the hippocampal CA1 region, the decrease in NMDA receptor channel activity should causally relate to the reduction of LTP in the ε1 mutant mice.

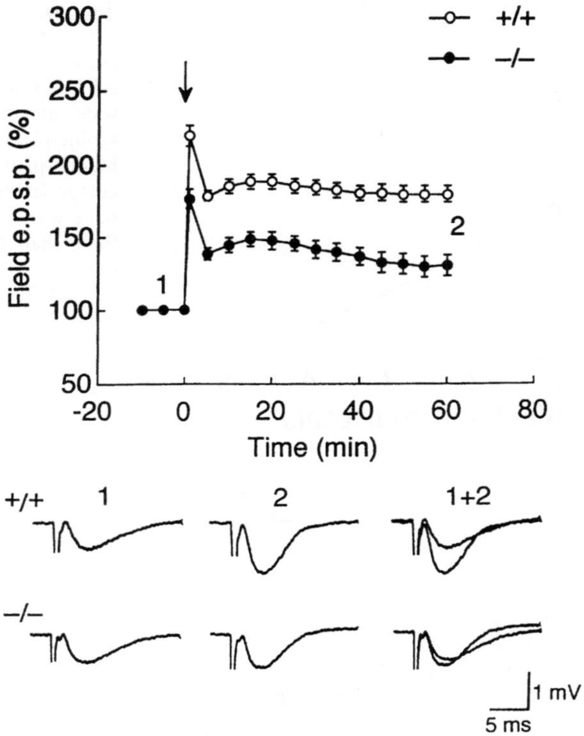

FIG. 5. Long-term potentiation (LTP) of hippocampal CA1 field *e.p.s.p.* (excitatory postsynaptic potential, EPSP) expressed as percentage of the mean before stimulation. A single slice was taken from one animal. Potentiation by tetanic stimulation (100 Hz for 1 s, 2 trains, 30-s interval; *arrow*) was recorded from 20 wild-type slices (*open circles*) and 17 mutant slices (*filled circles*)

We employed the Morris water-maze task to examine whether the ε1 mutant mice can learn to perform a complex spatial learning task [41]. As a control, mice were given a visible-platform task to evaluate basic abilities such as motivation, motor ability, and sight. The mutant mice performed the task as efficiently as the wild-type mice, except during the first block of trials (Fig. 6a). These data show that the ε1 mutant and wild-type mice have sufficient ability to perform the nonspatial learning water-maze task. Then, the spatial learning ability of the mutant mice was tested in the hidden-platform version of the Morris water-maze task. The wild-type mice quickly learned to locate the hidden platform. On the other hand, the mutant mice took a significantly longer time than the wild-type mice. The trained mice were further examined by a transfer test. The wild-type mice spent more time in the trained quadrant where the platform had been located during training than did the mutant mice. These findings suggest that the ε1 mutant mice perform a spatial learning task less well and less precisely than the wild-type mice.

In the current investigation, we have shown by gene targeting that the GluRε1 mutant mice with reduced hippocampal LTP exhibit a moderate defect in spatial learning. These results support the notion that NMDA receptor channel-dependent synaptic plasticity is the cellular basis of certain forms of learning.

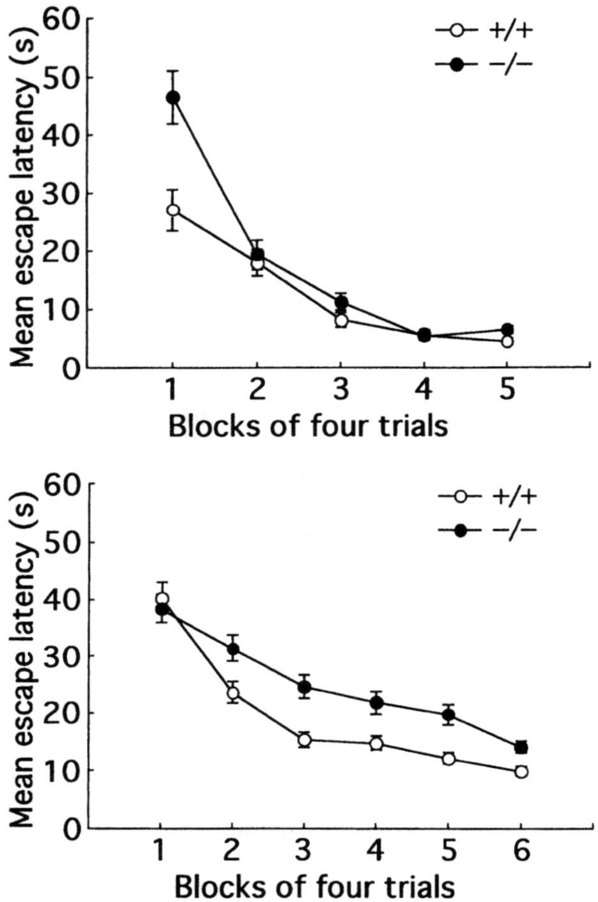

a

b

Fig. 6a,b. Mean escape latency of the wild-type (*open circles*) and mutant (*filled circles*) mice in the Morris water-maze task. **a** The visible-platform test. **b** The hidden-platform test. The escape latency of the mutant mice was significantly longer than that of the wild-type mice

Conclusion

We have identified more than a dozen subunits of the GluR channel by molecular cloning. These subunits can be classified into six subfamilies according to the amino acid sequence homology. Functional analysis of the cloned GluR channel subunits cDNAs shows that this classification, based purely on structural similarity, corresponds very well to the functional properties. We have demonstrated that NMDA receptor channels with high activity are formed by coexpression of the distantly related ε- and ζ-subunits. The four ε-subunits are highly homologous in primary structure, but are clearly distinct in distribution, functional properties, and regulation. The functional properties of the NMDA receptor channel are critically determined by the constituting ε-subunit, and thus the molecular diversity of the ε-subunit family underlies the functional heterogeneity of the NMDA receptor channel. Disruption of the ε1-subunit did not appreciably affect the growth and mating of the mice, but resulted in significant reduction of the NMDA receptor channel current and LTP at the hippocampal CA1 synapses. The ε1 mutant mice also showed a moderate deficiency in learning ability that was estimated by the Morris water-maze task. These results support the notion that NMDA receptor channel-dependent synaptic plasticity is the cellular basis of certain forms of learning.

Acknowledgments. We thank Drs. Hisashi Mori and Masahiko Watanabe, Ms. Etsuko Kushiya, and Ms. Rie Natsume for help in preparation of the manuscript. This work was supported in part by the Ministry of Education, Science and Culture of Japan, and the Uehara Memorial Foundation.

References

1. Mayer ML, Westbrook GL (1987) The physiology of excitatory amino acids in the vertebrate central nervous system. Prog Neurobiol 28:197–276
2. Monaghan DT, Bridges RJ, Cotman CW (1989) The excitatory amino acid receptors: their classes, pharmacology, and distinct properties in the function of the central nervous system. Annu Rev Pharmacol Toxicol 29:365–402
3. Sugiyama H, Ito I, Hirono C (1987) A new type of glutamate receptor linked to inositol phospholipid metabolism. Nature 325:531–533
4. Bliss TVP, Collingridge GL (1993) A synaptic model of memory: long-term potentiation in the hippocampus. Nature 361:31–39
5. Ito M (1989) Long-term depression. Annu Rev Neurosci 12:85–102
6. McDonald JW, Johnston MV (1990) Physiological and pathophysiological roles of excitatory amino acids during central nervous system development. Brain Res Rev 15:41–70
7. Choi DW (1988) Glutamate neurotoxicity and diseases of the nervous system. Neuron 1:623–634
8. Olney JW (1990) Excitotoxic amino acids and neuropsychiatric disorders. Annu Rev Pharmacol Toxicol 30:47–71
9. Hollmann M, O'Shea-Greenfield A, Rogers SW, Heinemann S (1989) Cloning by functional expression of a member of the glutamate receptor family. Nature 342:643–648
10. Sakimura K, Bujo H, Kushiya E, Araki K, Yamazaki M, Yamazaki M, Meguro H, Warashina A, Numa S, Mishina M (1990) Functional expression from cloned cDNAs of glutamate receptor species responsive to kainate and quisqualate. FEBS Lett 272:73–80
11. Morita T, Sakimura K, Kushiya E, Yamazaki M, Meguro H, Araki K, Abe T, Mori KJ, Mishina M (1992) Cloning and functional expression of a cDNA encoding the mouse β2 subunit of the kainate-selective glutamate receptor channel. Mol Brain Res 14:143–146
12. Sakimura K, Morita T, Kushiya E, Mishina M (1992) Primary structure and expression of the γ2 subunit of the glutamate receptor channel selective for kainate. Neuron 8:267–274
13. Yamazaki M, Araki K, Shibata A, Mishina M (1992) Molecular cloning of a cDNA encoding a novel member of the mouse glutamate receptor channel family. Biochem Biophys Res Commun 183:886–892
14. Araki K, Meguro H, Kushiya E, Takayama C, Inoue Y, Mishina M (1993) Selective expression of the glutamate receptor channel δ2 subunit in cerebellar Purkinje cells. Biochem Biophys Res Commun 197:1267–1276
15. Yamazaki M, Mori H, Araki K, Mori KJ, Mishina M (1992) Cloning, expression and modulation of a mouse NMDA receptor subunit. FEBS Lett 300:39–45
16. Meguro H, Mori H, Araki K, Kushiya E, Kutsuwada T, Yamazaki M, Kumanishi T, Arakawa M, Sakimura K, Mishina M (1992) Functional characterization of a heteromeric NMDA receptor channel expressed from cloned cDNAs. Nature 357:70–74
17. Kutsuwada T, Kashiwabuchi N, Mori H, Sakimura K, Kushiya E, Araki K, Meguro H, Masaki H, Kumanishi T, Arakawa M, Mishina M (1992) Molecular diversity of the NMDA receptor channel. Nature 358:36–41

18. Ikeda K, Nagasawa M, Mori H, Araki K, Sakimura K, Watanabe M, Inoue Y, Mishina M (1992) Cloning and expression of the ε4 subunit of the NMDA receptor channel. FEBS Lett 313:34–38
19. Noda M, Takahashi H, Tanabe T, Toyosato M, Kikyotani S, Furutani Y, Hirose T, Takashima H, Inayama S, Miyata T, Numa S (1983) Structural homology of *Torpedo californica* acetylcholine receptor subunits. Nature 302:528–532
20. Schofield PR, Darlison MG, Fujita N, Burt DR, Stephenson FA, Rodriguez H, Rhee LM, Ramachandran J, Reale V, Glencorse TA, Seeburg RH, Barnard EA (1987) Sequence and functional expression of the $GABA_A$ receptor shows a ligand-gated receptor superfamily. Nature 328:221–227
21. Grenningloh G, Rienitz A, Schmitt B, Methfessel C, Zensen M, Beyreuther K, Gundelfinger ED, Betz H (1987) The strychnine-binding subunit of the glycine receptor shows homology with nicotinic acetylcholine receptors. Nature 328:215–220
22. Mori H, Mishina M (1995) Structure and function of the NMDA receptor channel. Neuropharmacology 34:1219–1237
23. Hollmann M, Heinemann S (1994) Cloned glutamate receptors. Annu Rev Neurosci 17:31–108
24. Keinänen K, Wisden W, Sommer B, Werner P, Herb A, Verdoorn TA, Sakmann B, Seeburg PH (1990) A family of AMPA-selective glutamate receptors. Science 249:556–560
25. Egebjerg J, Bettler B, Hermans-Borgmeyer I, Heinemann S (1991) Cloning of a cDNA for a glutamate receptor subunit activated by kainate but not AMPA. Nature 351:745–748
26. Herb A, Burnashev N, Werner P, Sakmann B, Wisden W, Seeburg PH (1992) The KA-2 subunit of excitatory amino acid receptors shows widespread expression in brain and forms ion channels with distantly related subunits. Neuron 8:775–785
27. Bettler B, Boulter J, Hermans-Borgmeyer I, O'Shea-Greenfield A, Deneris ES, Moll C, Borgmeyer U, Hollmann M, Heinemann S (1990) Cloning of a novel glutamate receptor subunit, GluR5: expression in the nervous system during development. Neuron 5:583–595
28. Werner P, Voigt M, Keinänen K, Wisden W, Seeburg PH (1991) Cloning of a putative high-affinity kainate receptor expressed predominantly in hippocampal CA3 cells. Nature 351:742–744
29. Moriyoshi K, Masu M, Ishii T, Shigemoto R, Mizuno N, Nakanishi S (1991) Molecular cloning and characterization of the rat NMDA receptor. Nature 354:31–37
30. Yamakura T, Mori H, Masaki H, Shimoji K, Mishina M (1993) Different sensitivities of NMDA receptor channel subtypes to non-competitive antagonists. Neuroreport 4:687–690
31. Mori H, Yamakura T, Masaki H Mishina M (1993) Involvement of the carboxyl-terminal region in modulation by TPA of the NMDA receptor channel. Neuroreport 4:519–522
32. Watanabe M, Inoue Y, Sakimura K, Mishina M (1992) Developmental changes in distribution of NMDA receptor channel subunit mRNAs. Neuroreport 3:1138–1140
33. Watanabe M, Mishina M, Inoue Y (1994) Distinct spatiotemporal expressions of five NMDA receptor channel subunit mRNAs in the cerebellum. J Comp Neurol 343:513–519
34. Ben-Ari Y, Cherubini K, Krnjevic K (1988) Changes in voltage dependence of NMDA currents during development. Neurosci Lett 94:88–92
35. Kleckner NW, Dingledine R (1991) Regulation of hippocampal NMDA receptors by magnesium and glycine during development. Mol Brain Res 11:151–159
36. Hestrin S (1992) Developmental regulation of NMDA receptor-mediated synaptic currents at a central synapse. Nature 357:686–689
37. Mishina M, Sakimura K, Mori H, Kushiya E, Harabayashi M, Uchino S, Nagahari K (1991) A single amino acid residue determines the Ca^{2+} permeability of AMPA-selective glutamate receptor channels. Biochem Biophys Res Commun 180:813–821
38. Mori H, Masaki H, Yamakura T, Mishina M (1992) Identification by mutagenesis of a Mg^{2+}-block site of the NMDA receptor channel. Nature 358:673–675

39. Mishina M, Mori H, Araki K, Kushiya E, Meguro H, Kutsuwada T, Kashiwabuchi N, Ikeda K, Nagasawa M, Yamazaki M, Masaki H, Yamakura T, Morita T, Sakimura K (1993) Molecular and functional diversity of the NMDA receptor channel. Ann NY Acad Sci 707:136–152
40. Sakimura K, Kutsuwada T, Ito I, Manabe T, Takayama C, Kushiya E, Yagi T, Aizawa S, Inoue Y, Sugiyama H, Mishina M (1995) Reduced hippocampal LTP and spatial learning in mice lacking NMDA receptor ε1 subunit. Nature 373:151–155
41. Morris RGM, Anderson E, Lynch GS, Baudry M (1986) Selective impairment of learning and blockade of long-term potentiation by an *N*-methyl-D-aspartate receptor antagonist, AP5. Nature 319:774–776

Metabotropic Glutamate Responses and the Intracellular Mechanisms

NORIO AKAIKE

Summary. Glutamate and its analogs are the predominant excitatory neurotransmitters in the central nervous system (CNS). They are known to be critically involved in physiological and pathological processes such as learning, memory, and neuronal degeneration. Based on electrophysiological and pharmacological studies, the glutamate receptors are classified into (1) ionotropic receptors (iGluR), which are directly coupled with ionic channels and are subdivided into α-amino-3-hydroxy-5-methyl-4-isoxazolepropionate/kainate (AMPA/KA) and *N*-methyl-D-aspartate (NMDA) receptors, and (2) metabotropic receptors (mGluR), which are coupled to either inositol-1,4,5-triphosphate (IP_3)/Ca^{2+} or cyclic adenosine monophosphate (cAMP) pathways for signal transduction. The aim of the study was to clearly identify the electrophysiological responses to mGluR stimulation in CNS neurons and to compare them in different brain areas. To circumvent these problems, we used acutely dissociated rat CNS neurons and a quick drug-application system termed the Y-tube method. Detailed electrophysiological analysis was performed with the nystatin perforated patch recording mode, which prevents the washout of soluble intracellular substances during recording. Using this novel technique, mGluR responses could be recorded continuously for more than 1 h. We characterized the mGluR responses in seven regions of the rat brain. The responses were either activation of Ca^{2+}-activated K^+ current ($I_{K(Ca)}$), M current (I_M) suppression, leakage K^+ current suppression, or a combination of these, depending on the brain regions to be studied. These results on the regional difference along with the antagonist studies indicate that there are multiple types of mGluR. Both $I_{K(Ca)}$ and I_M are known to regulate the firing patterns of neurons and the amplitude of the postsynaptic potential. Moreover, the threshold concentration for mGluR response was 10–100 fold lower than that for iGluR response, indicating a wider range of operation. On the basis of these facts, we propose that mGluR is a finely tuned modulator of membrane excitability in the CNS.

Key words. Rat brain—Dissociated neuron—Nystatin perforated patch—Glutamate—Metabotropic receptor—IP_3—Ca^{2+}—Ca^{2+}-activated K^+ current ($I_{K(Ca)}$)—M current (I_M)—Leakage K^+ current

---

Department of Physiology, Kyushu University Faculty of Medicine, 3-1-1 Maidashi, Higashi-ku, Fukuoka 812-82, Japan

Introduction

Glutamate (Glu) acts as a major and important excitatory neurotransmitter at many synapses in the mammalian central nervous system (CNS). Glu receptors have been classified into two categories, ionotropic and metabotropic receptors (iGluR and mGluR, respectively) [1]. iGluR is a complex with a nonspecific cation channel, and its pharmacological and biochemical characteristics and physiological functions have been well investigated [2]. On the other hand, there is only fragmentary information on mGluR-induced responses mediated by guanine nucleotide-binding proteins (G proteins) and second-messenger systems.

mGluR has been revealed to have at least seven different subtypes, termed mGluR1–mGluR7, which have been further subdivided into three subgroups [1,3,4]. The first subgroup, mGluR1 and mGluR5, are coupled to stimulation of phosphatidylinositol (PI) hydrolysis and the subsequent mobilization of Ca^{2+} from intracellular stores. The second subgroup, mGluR2–mGluR3, and the third group, mGluR4, mGluR6, and mGluR7, are linked to the inhibition of the cyclic adenosino monophosphate (cAMP) cascade. Pharmacological and molecular biological studies also suggested the regional difference of the localization of these mGluRs in the brain [1,5]. Therefore, it is of great interest to discover what functional roles these mGluR subtypes might play in the various regions of the mammalian CNS. The electrophysiological investigation of direct mGluR stimulation of dissociated neurons in various CNS region seems to be important and useful to understand the signal transduction and detailed physiological and pharmacological characteristics.

Regional Differences of mGluR Responses in Rat CNS Neurons

To survey the differences in mGluR responses in the neurons acutely dissociated from various brain regions in the rat, the mGluR responses were measured in the caudate putamen (CP), nucleus of the solitarius (NTS), cerebellar Purkinje neurons (CPJ), granule cells in the main olfactory bulb (mOB), hippocampal CA1 and CA3 pyramidal neurons (CA1, CA3), and granule cells in the dentate gyrus (DG). As shown in Fig. 1, in most brain regions mGluR activation by Glu in the presence of 6-cyano-7-nitroquinoxaline-2,3-dione (CNQX) evoked a slow inward current accompanied by a membrane conductance decrease at a holding potential (V_H) of -44 mV. In the CP, an inward current was observed in most neurons (>90%) and showed a larger amplitude than in other regions. In mOB, CPJ, and NTS neurons, Glu also elicited inward currents. These mGluR-induced inward currents reached a steady state in 20–30 s and gradually recovered within 2–3 min after the washout of the agonist. In some neurons of CA1, CA3, and DG, however, an additional transient outward current with conductance increase was observed, and this current was followed by a slow inward current with decreased conductance.

Figure 2 shows the ratio of mGluR response to that of iGluR in individual brain regions. As the value of iGluR response, the maximal amplitude of transient peak inward current during $10^{-4} M$ Glu application was used, and the maximal amplitude of the slow inward current during Glu application in the presence of $10^{-4} M$ CNQX was used for the mGluR response. The inward mGluR/iGuR response ratio was largest in the CP neuron (0.448 ± 0.138), and the rank order of the ratio was CP > mOB > CA3 > CPJ > NTS > CA1.

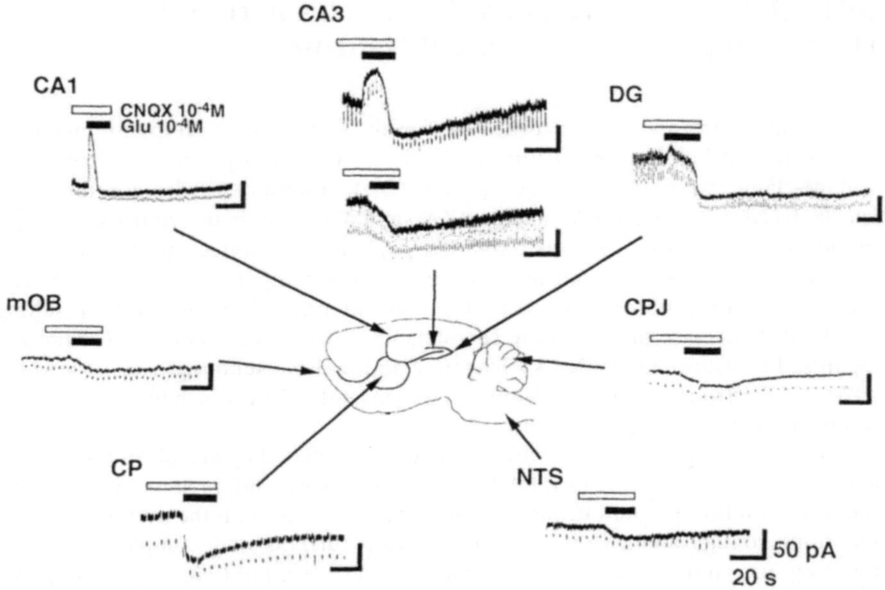

Fig. 1. Regional difference of metabotropic glutamate receptor (mGluR) responses in the rat brain. The responses induced 10^{-4} *M* Glu in the presence of 10^{-4} *M* 6-cyano-7-nitro-quinoxaline-2,3-dione (CNQX) under voltage-clamp conditions [holding potential (V_H), −44 mV]. *Center panel*, the submediant plane slice of rat brain. *CA1* and *CA3*, regions of hippocampus; *DG*, dentate gyrus; *mOB*, granule cells in the main olfactory bulb; *CPJ*, cerebellar Purkinje neurons; *CP*, caudate putamen; *NTS*, nucleus of solitarius. Traces are typical examples in each region. Hyperpolarizing voltage pulses of about 10 mV and 350–600 ms in duration were continuously applied every 3–6 s

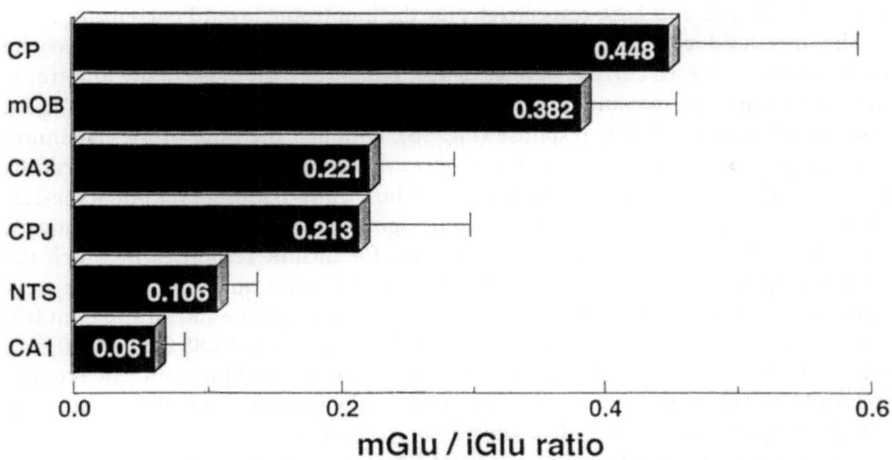

Fig. 2. Regional difference of mGluR/iGluR response ratio. The ionotropic receptor (iGluR) response is the ionotropic value, which is the maximal amplitude of transient peak inward current under Glu (10^{-4} *M*) application. The mGluR response is the metabotropic value, the maximal amplitude of the slow inward current under Glu (10^{-4} *M*) and CNQX (10^{-4} *M*) coapplication. Each value indicates mean ± SEM for 5–8 cells examined

mGluR-Induced Transient Outward Current in Hippocampal CA1 Pyramidal Neurons

When ionic currents were recorded in freshly dissociated CA1 pyramidal neurons of the rat hippocampus using conventional whole-cell patch recording under voltage-clamp condition, Glu, quisqualate (QA), α-amino-3-hydroxy-5-methyl-4-isoxazoleproprionate (AMPA), kainate (KA), and *N*-methyl-D-aspartate (NMDA) induced rapid inward currents. In the nystatin perforated patch recording configuration, however, a transient outward current component overrode the early phase of the steady-state inward current component induced by Glu and QA, resulting in a further increase in membrane conductance. This outward current was not antagonized by iGluR antagonists such as DL-2-amino-5-phosphonopentanoic acid (D-AP5), 5,7-dinitro quinoxaline-2,3-dione (DNQX), and CNQX, which blocked only the inward currents mediated by iGluR.

The results suggest that this outward current is mediated by mGluR in the hippocampal CA1 pyramidal neurons. NMDA, aspartate (Asp), and, in some neurons, KA induced both inward and outward current when recorded with the perforated patch recording configuration. However, such outward currents were completely antagonized by DL-2-amino-5-phosphonopentanoic acid (D-AP5) or CNQX, indicating that the outward currents were presumably activated by Ca^{2+} passing through the nonspecific cation channels gated by the NMDA and type II KA receptor. Interestingly, Glu and QA induced the metabotropic outward currents at lower concentrations, and an additional ionotropic inward current was elicited at higher concentrations (Fig. 3a). Concentration–response relationships for both outward and inward currents induced by Glu and QA showed a sigmoidal increase with increasing concentrations. When the peak amplitudes of both inward and outward currents were normalized to the ionotropic peak current amplitude induced by $10^{-5} M$ QA, the threshold concentration, EC_{50} value, apparent Hill coefficient, and maximum response were $10^{-8} M$, $1.1 \times 10^{-7} M$, 2.29 and 0.78, respectively, for the metabotropic outward current, and $10^{-7} M$, $1.5 \times 10^{-5} M$, 0.82 and 2.29, respectively, for the ionotropic inward current.

The threshold concentration and effective concentration (EC_{50}) value of the metabotropic outward current were 10 to 100 fold lower than those of the ionotropic inward current. The maximum current amplitude of the mGluR response was about one-third that of the iGluR response (Fig. 3b), although the ratio of the maximum amplitude of mGluR to that of iGluR responses varied from cell to cell. Similar results were obtained by the application of Glu. *trans*-1-Aminocyclo-pentane-1,3-dicarboxylate (*t*-ACPD), a selective mGluR agonist [6], also induced an outward current. In the current–response relationships for mGluR responses in which the mGluR responses induced by Glu, QA, and *t*-ACPD were normalized to the peak outward current amplitude induced by $10^{-4} M$ Glu, the EC_{50} value and an apparent Hill coefficient were $1.8 \times 10^{-6} M$ and 2.22 for Glu, $1.5 \times 10^{-4} M$ and 1.88 for *t*-ACPD, and $1.1 \times 10^{-7} M$ and 2.29 for QA. The relative maximum values of these mGluR agonist-induced currents were very similar. The threshold concentrations of QA, Glu, and *t*-ACPD responses were 10^{-8}, 10^{-7}, and $10^{-5} M$, respectively.

Neurons were perfused with external and internal solutions containing K^+ at concentrations of 5 and 150 mM, respectively. To elucidate the current–voltage (I–V) relationship of mGluR response, a depolarizing triangular voltage ramp of 20–30 mV, followed by a mirror-image hyperpolarizing ramp, was applied before and during Glu

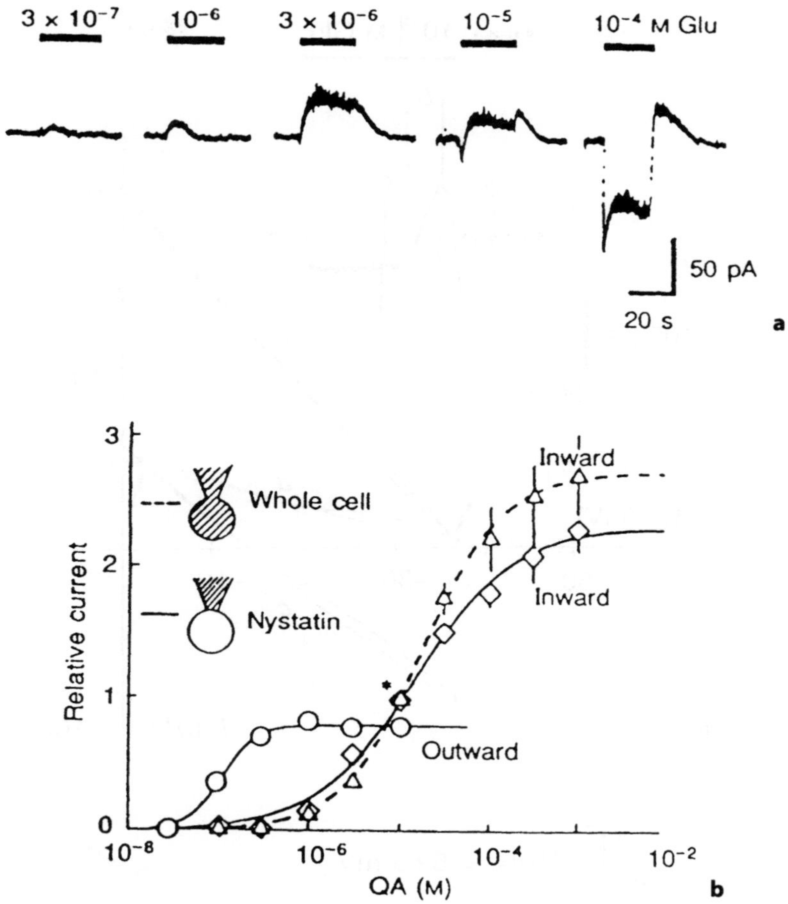

FIG. 3a,b. Excitatory amino acid- (EAA-) induced currents. **a** Glu induced both mGluR and iGluR responses at a V_H of −20 mV. **b** Concentration–response curve of iGluR (*diamonds*) and mGluR (*circles*) responses induced by (QA) in the nystatin perforated-patch recording configuration. Quisqualate (QA)-induced iGluR response with the conventional whole-cell patch recording (triangles) is also shown. All responses were normalized to the peak inward current induced by 10^{-5} *M* QA (*). Each point is the average of four experiments; vertical bars show ± SEM

application (Fig. 4a, inset). The intercept of the curves obtained before Glu application (*a*) and during a steady-state inward current (*c*) gave the reversal potential of the iGluR response (E_{iGlu}) while the intercept of the two I-V curves obtained before Glu application (*a*) and near the peak of the Glu-induced outward current (*b*) gave the reversal potential of a composite response ($E_{mGlu+iGlu}$) consisting of both mGluR and iGluR responses. The measured E_{iGlu} was +2.75 ± 1.91 mV, which was close to the theoretical equilibrium potential (E_{cation}) of +1.3 mV. This result suggests that the steady-state inward component of the Glu response was caused by the activation of large cation channels coupled with iGluR, as reported previously [2].

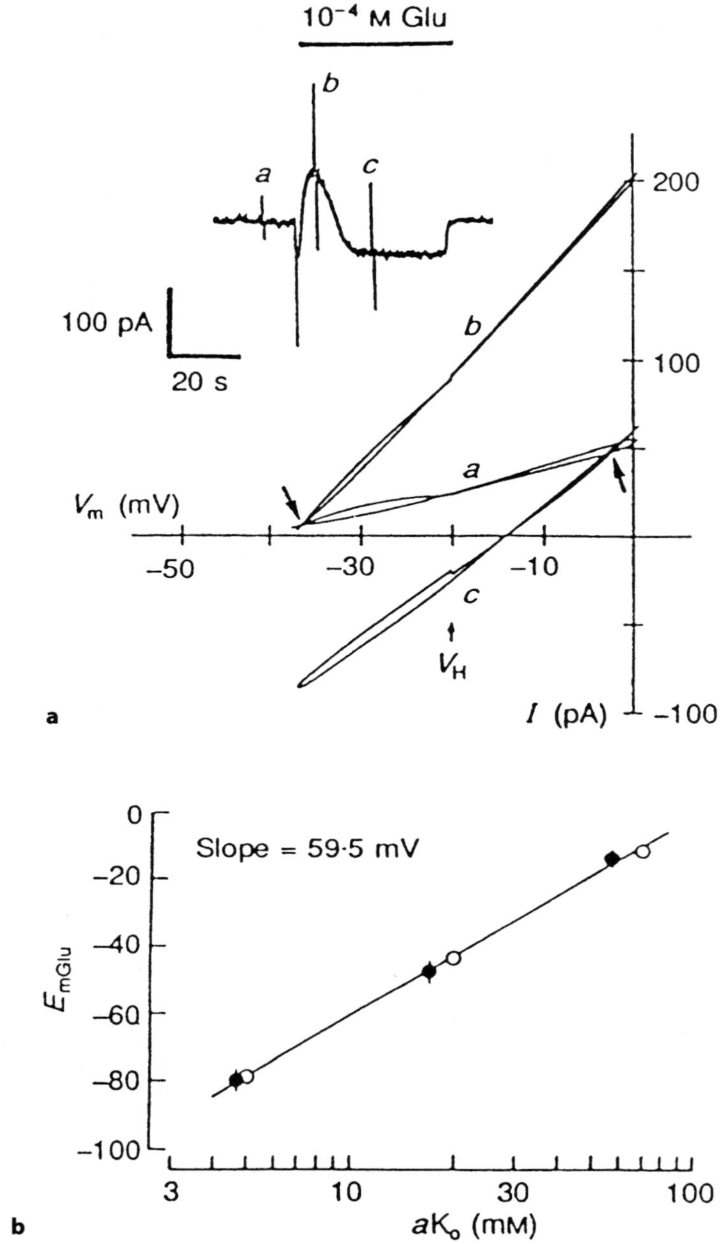

FIG. 4a,b. Current–voltage (*I–V*) relationship for Glu response assessed with the nystatin perforated patch recording configuration. **a** I–V curves recorded before Glu ($10^{-4} M$) application (*a*), at the peak of outward current (*b*), and at the steady-state inward current (*c*). V_H, −20 mV. **b** Relationship between the mGluR response (E_{mGlu}) and extracellular potassium activity (aK_o). *Open circles*, theoretical K^+ equilibrium potential (E_K) values; *Solid circles*, measured E_{mGlu}. Each point is the average of four experiments

On the other hand, $E_{mGlu+iGlu}$ was -18.87 ± 4.53 mV. This value was between E_{cation} and the theoretical K^+ equilibrium potential (E_K) of -80.5 mV, suggesting that the reversal potential of the mGluR response (E_{mGlu}) is close to E_K. To measure the pure E_{mGlu}, therefore, I-V relationships were made in the external solution containing iGluR antagonists at various concentrations of extracellular K^+ ($[K^+]_o$). The intracellular K^+ concentration ($[K^+]_i$) was fixed at 150 m*M*. A hyperpolarizing triangular voltage ramp of 40 mV was applied before Glu application and at the time when the mGluR response reached its peak. In standard external solution with 5 m*M* K^+, E_{mGlu} was -79.4 ± 2.21 mV, which was close to E_K (-80.45 mV). When $[K^+]_o$ was increased, E_{mGlu} was shifted to depolarizing potentials. The E_{mGlu} values at 20 and 70 m*M* $[K^+]_o$ were -46.73 ± 2.89 and -13.24 ± 1.67 mV, respectively, which were also close to the respective E_K values (-46.92 and -17.9 mV). The plot of E_{mGlu} against the $[K^+]_o$ activity (aK_o) gives the slope of 59.5 mV for a tenfold shift in $[K^+]_o$ (Fig. 4b), indicating that the mGluR-operated current is passing through K^+ channels.

The conventional whole-cell recording with a patch pipette filled with internal solution containing 2 m*M* ethyleneglycoltetraacetic acid (EGTA) failed to detect the mGluR response. The mGluR response also disappeared when the mode of patch recording was switched from the perforated-patch mode to the conventional whole-cell mode in the same neuron. In the nystatin perforated-patch recording, the mGluR response was recorded even in Ca^{2+}-free external solution, although a second application of Glu 2 min after the first application did not induce any mGluR response. This interval of 2 min was long enough for the mGluR response to recover almost completely from desensitization. In addition, even in Ca^{2+}-free external solution, $[Ca^{2+}]_i$ increased with the first application of either Glu or QA but not with the iGluR agonist AMPA. The results indicate, first, that the mGluR-induced outward current is a Ca^{2+}-dependent K^+ current ($I_{K(Ca)}$) activated with Ca^{2+} that may be released from intracellular Ca^{2+} stores, and second, that the Ca^{2+} stores are rapidly depleted in the absence of extracellular Ca^{2+}.

Two different intracellular Ca^{2+} stores have been reported in neurons [7]; one is an inositol 1,4,5-triphosphate- (IP_3-) induced Ca^{2+} release (IICR) pool that is inhibited by heparin, protein kinase A, H^+, and Mg^{2+}, and another is a Ca^{2+}-induced Ca^{2+} release (CICR) pool, which is stimulated by caffeine and Ca^{2+} of low concentrations, but inhibited by high concentrations of Ca^{2+}, ruthenium red, calmodulin, H^+, and Mg^{2+}. Pretreatment with either acetylcholine (ACh) or caffeine also markedly suppressed the mGluR-induced outward current. We have already reported that ACh releases Ca^{2+} from IICR in the same preparation [8]. Ryanodine, which is known to lock the Ca^{2+} release channel on the CICR membrane in the "open state" [7], also showed use-dependent and irreversible suppression of the mGluR response. These results suggest that both IICR and CICR are necessary for mGluR responses.

To allow intracellular application of modulators and to study their effects, the rundown of mGluR responses under the conventional whole-cell recording mode was partly prevented by adding 0.5 m*M* guanosine triphosphate (GTP), 5 m*M* ATP, and 6 m*M* Mg^{2-} to the internal solution. Changes in the time-course of rundown were then used to assess the positive effects of the modulators. The mGluR response increased at first, and then time-dependently decreased. Relative current amplitude at 30 min was 0.46 ± 0.24. GDP-β-S, a nonhydrolyzable analog of guanose diphosphate (GDP), was used for blocking the G protein-mediated effects of neurotransmitters [9]. Inclusion of GDP-β-S in the internal solution accelerated rundown, and the response disappeared completely 18 min after the initial recording, indicating the involvement of G proteins in the mGluR response.

Neomycin (an inhibitor of phospholipase C, PLC) and heparin (a competitive antagonist of the IP_3 receptor) facilitated rundown. In addition, the mGluR response was suppressed during the activation of $I_{K(Ca)}$ induced by IP_3 in the internal solution. Pretreatment with staurosporine, an inhibitor of PKC [10], at $10^{-7} M$ for 4 min did not affect the mGluR response. In the same preparation, thyrotropin-releasing hormone (TRH) induced a K^+ current that is dependent on PKC activity [11]. Four minutes of pretreatment with staurosporine was effective in inhibiting the TRH-elicited current, with an IC_{50} value of $2.45 \times 10^{-8} M$. Thus, the absence of modulation of mGluR response by staurosporine at $10^{-7} M$ suggests the lack of PKC involvement. The results clearly indicate the involvement of the IP_3-Ca^{2+} system but not that of PKC in the mGluR response.

Because previous reports on biochemical properties have suggested that pertussis toxin- (IAP-) sensitive G proteins couple to mGluR [1,12,13], the effect of IAP was examined using the perforated-patch recording. The dissociated pyramidal neurons were incubated in standard external solution with or without IAP (200 μg ml^{-1}) for 6–

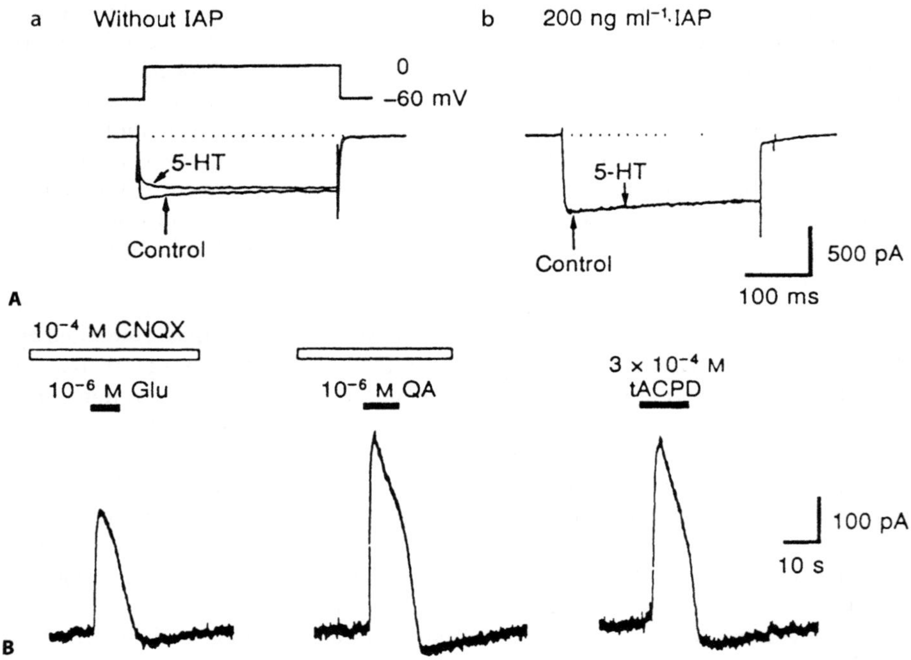

Fig. 5A,B. Effects of pertussis toxin (IAP) on mGluR response. **A** Effect of IAP treatment on the inhibition of high-voltage-activated Ba^{2+} current (HVA I_{Ba}) by serotonin (5-HT). Neuronal HVA I_{Ba} was induced by a step depolarization from −60 to 0 mV after 6–8 h of cell dissociation in the presence (*b*) or absence (*a*) of IAP. HVA I_{Ba} was inhibited by $10^{-5} M$ 5-HT in control neurons without IAP treatment, but the inhibitory action of 5-HT on HVA I_{Ba} was never observed in IAP-treated neurons. Neurons were pretreated with 5-HT for more than 2 min. *Control*, HVA I_{Ba} just before 5-HT application. **B** Glu, QA, and *trans*-1-aminocyclo-pentane-1,3-dicarboxylate (*t*-ACPD) induced mGluR responses in neurons treated with IAP. V_H, −40 mV. **A** and **B** were recorded from the same neuron

8h [8] at room temperature. To ascertain that the treatment with IAP was effective in these dissociated neurons, the effect of serotonin (5-HT) on the high-voltage-activated (HVA) Ca^{2+} channels was also examined. Ca^{2+} in the external solution was replaced by Ba^{2+} to avoid the Ca^{2+} release from CICR and the activation of Ca^{2+}-dependent K^+ (K^+_{Ca}) channels. In the control neurons without IAP treatment, the Ba^{2+} current passing through the HVA Ca^{2+} channel (HVA I_{Ba}) was suppressed by the application of 5-HT, with its prominent effect on the activation phase (Fig. 5A*a*). However, 5-HT did not suppress HVA I_{Ba} in neurons treated with IAP (Fig. 5A*b*), substantiating the validity of our protocol for IAP treatment. Glu, QA, and *t*-ACPD could induce large mGluR outward currents even in the IAP-treated neurons (Fig. 5b), indicating that IAP-insensitive G proteins are coupled with the mGluR.

Three different types of K^+_{Ca} channels have been classified [14]. One class (high-conductance K^+_{Ca} channels) has a large single channel conductance of 100–250pS and is blocked by tetraethylammonium (TEA), quinine, Ba^{2+}, charybdotoxin, and iberiotoxin (IbTX). A second class (small-conductance K^+_{Ca} channels) has a small K^+ conductance between 6 and 14pS, and is blocked by TEA, quinine, and apamin. The third class (intermediate-conductance K^+_{Ca} channels) has an intermediate K^+ conductance (18–50pS), and is blocked by TEA, quinine, Cs^+, and Ba^{2+}. Characteristically, this type of K^+_{Ca} channel is insensitive to apamin and IbTX. Therefore these K^+ channel blockers were used for assessing the type of K^+_{Ca} channel responsible for mGluR in the hippocampal CA1 pyramidal neurons. A general K^+ channel blocker, TEA, blocked the mGluR-mediated $I_{K(Ca)}$, but apamin and IbTX had no effect, suggesting that intermediate-conductance K^+_{Ca} channels are involved in mGluR responses.

Figure 6 shows the intracellular signaling pathways for mGluR-induced transient outward current in acutely dissociated rat hippocampal CA1 pyramidal neurons: the mGluR, which is insensitive to both AP3 and AP4, couples with IAP-insensitive G protein (possibly G_q) and stimulates IP_3 formation through the activation of PLC. IP_3 stimulates Ca^{2+} release from IICR, and the elevated $[Ca^{2+}]_i$ promotes Ca^{2+} release from CICR, eventually activating the intermediate-conductance K^+_{Ca} channel (IK channel in Fig. 6) on the cytoplasmic membrane. The relative agonist affinity was QA > Glu > t-ACPD. This order seems to be common for mGluRs coupled with IP_3 formation [15–19].

mGluR-Induced Slow Inward Current in Caudate Putamen (CP) Neurons

In neurons freshly dissociated from the rat CP, 10^{-4} Glu induced a fast ionic current followed by a slow inward current. This slow inward current was produced even in the presence of CNQX ($10^{-4}M$), a specific iGluR antagonist, with decreasing conductance (Fig. 7a), indicating that this inward current was mediated by mGluR. Figure 3b summarizes the concentration–response relationships for iGluR and mGluR responses. When the peak amplitudes of iGluR and mGluR responses were normalized to the iGluR peak current amplitude induced by $10^{-4}M$ Glu, the threshold concentration, the half-maximal effective concentration (EC_{50}), and Hill coefficient were $\sim 3 \times 10^{-6}M$, $8.3 \times 10^{-5}M$, and 0.80 for the iGluR responses, and $\sim 3 \times 10^{-7}M$, $3.01 \times 10^{-6}M$, and 1.1 for the mGluR responses. The threshold concentration and EC_{50} value of the mGluR-induced inward current were 10- to 30 fold lower than those of the iGluR currents.

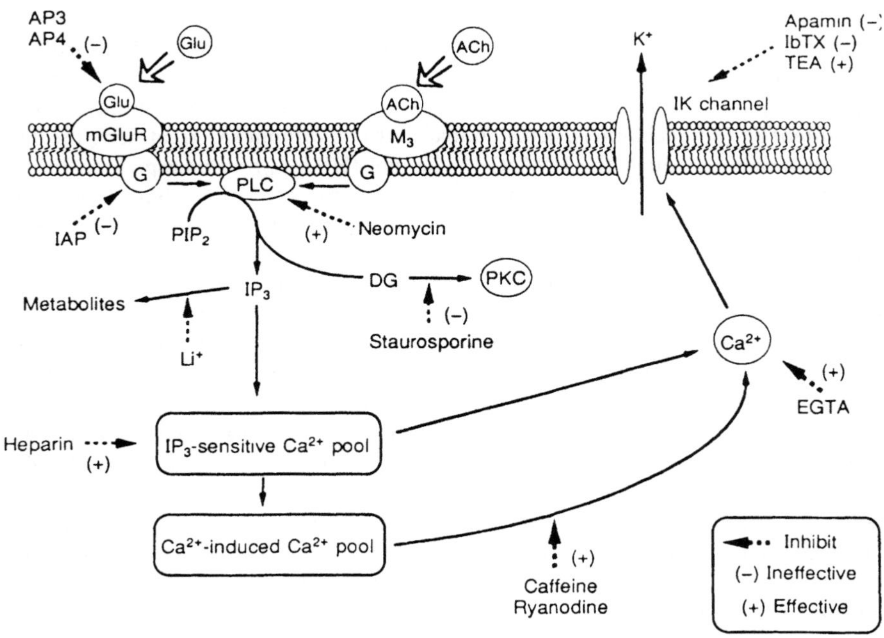

Fig. 6. Schematic diagram of possible intracellular signal transduction pathway via mGluR in acutely dissociated CA1 pyramidal neurons of the rat hippocampus. ACh, acetylcholine; *PKC*, protein kinase C; *TEA*, tetracthylammonium

QA, (1S,3R)-ACPD, and L-CCG-1 are known as the agonists of mGluR. Especially, (2S,3S,4S)-α-carboxycyclopropyl-glycine (L-CCG-1) is a restricted Glu analog and its pharmacological potency for mGluR1 is different from that for mGluR2 [20]. When these three agonists were applied to CP neurons, an inward current was elicited. The response was similar to the response induced by Glu application in the presence of CNQX. As assessed by EC_{50} values, the apparent affinity of these agonists was of the order QA > Glu > L-CCG-1 > (1S,3R)-ACPD for mGluR response. This order was the same for mGluR1 and mGluR5. L-CCG-1 did not induce any current below $10^{-6}M$, which concentration had submaximum effect for inhibition of forskoline-stimulated cAMP formation mediated by mGluR2, but did not affect PI hydrolysis mediated by mGluR1 [20].

The ions responsible for this mGluR current component were determined by the current–voltage (I–V) relationship using the ramp-wave method. When the patch pipette (internal) solution and external solution contained 150 and 5 m*M* K^+, respectively, *t*-ACPD ($10^{-4}M$) elicited an inward current and the I-V relationship slope decreased during response. The reversal potential was -79.9 ± 2.9 mV, which was close to the E_K of -82 mV (Fig. 8A). The slope of the reversal potential for a tenfold change in extracellular potassium activity (aK_o) was 56.4 mV. These results indicated that K^+ was the dominant charge carrier for the mGluR-mediated inward current.

The nature of ion channels about the inward current observed in the CP neurons was examined. Step hyperpolarizing pulses were given from aV_H of -30 mV, changing amplitudes in 10-mV steps from 10 to 60 mV, and 900 ms in duration. The pulse elicited an instantaneous current before *t*-ACPD application without being

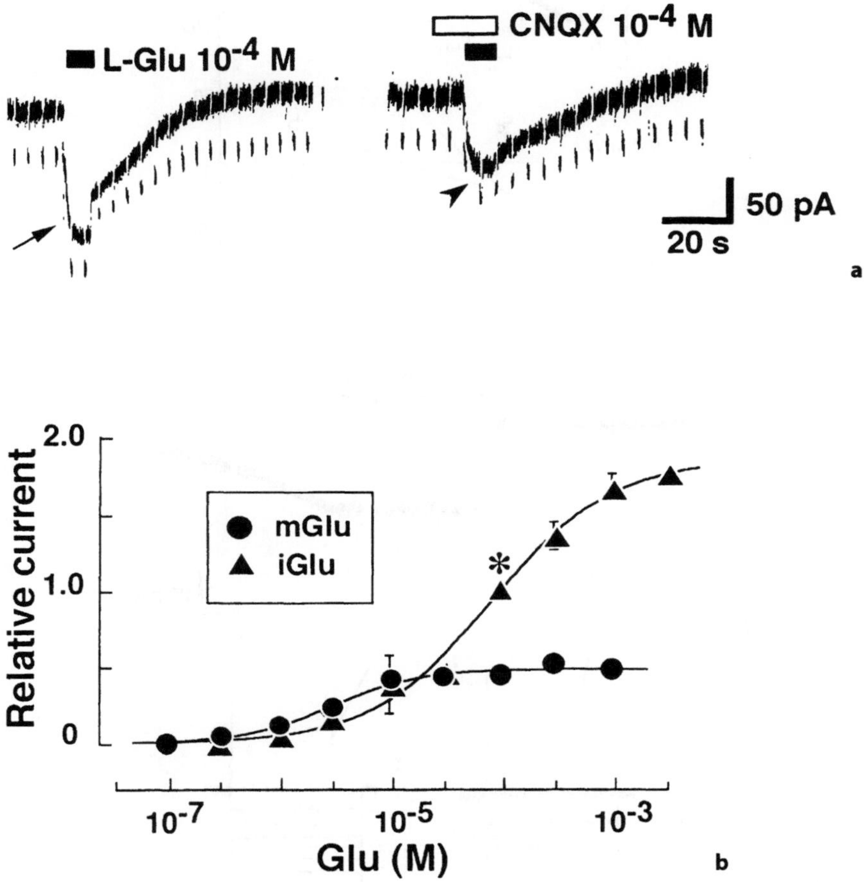

Fig. 7a,b. mGluR-Induced slow inward current. **a** *Left panel*, current evoked by $10^{-4}M$ Glu; *right panel*, current evoked by $10^{-4}M$ Glu in the presence of $10^{-4}M$ CNQX at a V_H of −44 mV. Hyperpolarizing voltage pulses of 10 mV and 600 ms in duration were applied every 6 s to monitor the changes in membrane conductance. The mGluR-induced inward current was accompanied by an decrease in membrane conductance. **b** Concentration–response curves for iGluR (*triangles*) and mGluR (*circles*) responses to Glu. Current amplitudes were normalized to that of the iGluR response induced by $10^{-4}M$ Glu (see symbol marked with *asterisk*). Each point is mean of three to six experiments; a vertical bar indicates ± SEM where this exceeded the size of the symbol

following by slow relaxations and voltage dependency. That current was markedly suppressed during inward responses evoked by *t*-ACPD ($3 \times 10^{-4}M$) (Fig. 8B,C). The results suggest that suppression of leakage K^+ current mainly underlies the inward currents induced by mGluR agonists of the CP neurons, but voltage- and time-dependent K^+ conductance (M current, abbreviated as I_M) has little influence on that response.

DL-AP3 and DL-AP4 are the classical antagonists of mGluR, but in our preparation, the (1S,3R)-ACPD-induced response was not affected by these substances. α-Methyl-

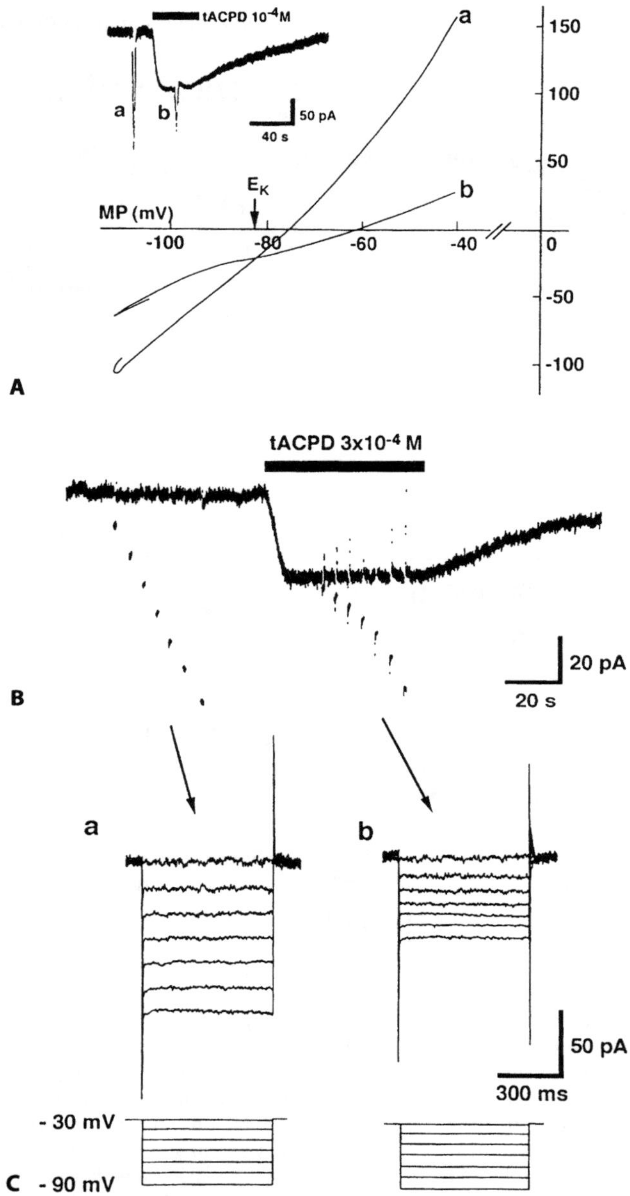

FIG. 8A–C. Suppression of leakage K^+ current by mGluR stimulation. **A** Current–voltage (I–V) relationships for the mGluR-mediated response. *Inset*, protocol for ramp-wave command. V_H, −44 mV; hyperpolarizing ramp commands from −44 to −125 mV of 3.3 s in duration were applied two times before 10^{-4} *M t*-ACPD treatment (*a*) and during inward response induced by *t*-ACPD (*b*). $[K^+]_i$ and $[K^+]_o$, 150 and 5 m*M*, respectively. E_K and *MP*, theoretical K^+ equilibrium potential and membrane potential, respectively. **B** mGluR activation by (1S, 3R)-ACPD suppressed a leak K^+ conductance. The step hyperpolarizing pulses were given from V_H of −30 mV, changing amplitudes in 10-mV steps from 10 to 60 mV and 900 ms in duration before agonist application and during inward response induced by *t*-ACPD. **C** The current elicited by step pulses does not show time- and voltage-dependency before *t*-ACPD application (*a*) and during the inward response evoked by *t*-ACPD (*b*).

4-carboxyphenyl-glycine (MCPG) is a new type of antagonist that selectively antagonized the (1S,3R)-ACPD-induced response, and the profile of this action appeared consistent with full competitive antagonism [21]. In this preparation, MCPG (10^{-3} *M*) did not induce any current by itself. After pretreatment with this antagonist for 1 min, however, the response produced by (1S, 3R)-ACPD (10^{-4} M) was inhibited.

The external application of Glu (10^{-4} *M*) in the presence of CNQX (10^{-4} *M*) slowly induced a membrane depolarization and elicited spontaneous action potentials. The excitability ceased gradually after the washout of Glu and CNQX. The application of 10^{-4} *M* TTX blocked the action potential firings but did not inhibit the depolarizing effect induced by Glu. The (1S, 3R)-ACPD (3×10^{-4} *M*) also induced depolarization and spontaneous action potentials (Fig. 9).

It was reported previously that activation of mGluR by *t*-ACPD depressed both an $I_{K(Ca)}$ and a voltage-gated K^+ current; the latter was independent of changes in intracellular Ca^{2+} using hippocampal slice cultures. However, a recent study using the same preparations has reported that in these conditions (whole-cell recording; V_H = −70 mV in the absence of cAMP in pipette solution), (1S, 3R)-ACPD and methacholine (MCh), acting cholinergic muscarinic receptors, evoked inward currents because of suppression of a leak K^+ conductance coupling to G protein [22]. Our group reported that application of mGluR agonists evoked an inward current in CA3 dissociated pyramidal neurons that was caused by suppression of both the leakage K^+ current and the M current (I_M). In rat cerebellar Purkinje cells (slice cultures), *t*-ACPD also induced a transient inward current, but this inward current was associated with an increase in apparent input conductance of cell membrane and with a rise in cytosolic Ca^{2+} concentration [23]. Thus, they suggested that this inward current was generated by Ca^{2+}-activated nonspecific cation channels or an electrogenic Na^+/Ca^{2+} exchanger.

In addition, in dissociated pyramidal cells of CA1 and CA3 in hippocampus, an outward current was caused by the Ca^{2+}-activated K^+ channel [24]. In neurons of the basolateral amygdala slices, activation of postsynaptic mGluRs resulted in membrane hyperpolarization associated with an increase in membrane input resistance. This hyperpolarization was suggested to be mediated by G protein and induced by activation of a TEA-sensitive, Ca^{2+}-dependent K^+ conductance [25].

In this study, the inward currents of the CP neurons observed under mGluR agonist application did not have voltage dependency, and their reversal potential was closed

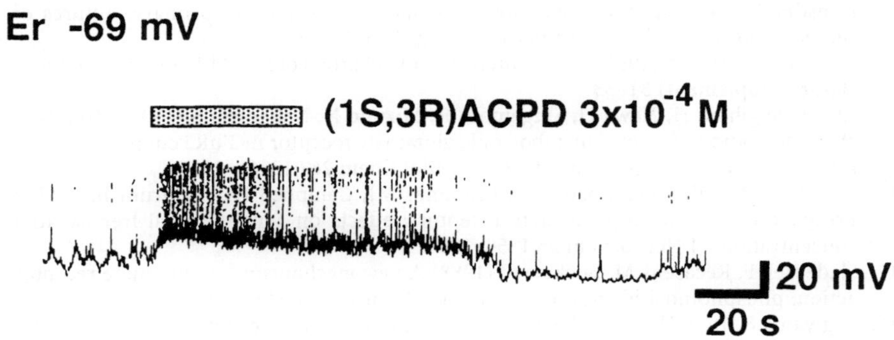

Fig. 9. A representative response evoked by mGluR activation under the current-clamp mode. (1S, 3R)-ACPD, 3×10^{-4} *M*; resting potential (E_r), −69 mV

to the E_K. Also, according to the step pulse study, I_M cannot be detected plainly in the CP neurons. The inward currents of this cell are thus suggested to be caused mainly by suppression of the leakage K^+ current. These results suggest that mGluR activation acts on some second-messenger pathways and drives different effectors in various regions in the CNS.

References

1. Nakanishi S (1992) Molecular diversity of glutamate receptors and implications for brain function. Science 258:597–603
2. Collingridge GL, Lester RAJ (1989) Excitatory amino acid receptors in the vertebrate central nervous system. Pharmacol Rev 40:143–210
3. Masu M, Tanabe Y, Tsuchida K, Shigemoto R, Nakanishi S (1991) Sequence and expression of a metabotropic glutamate receptor. Nature 349:760–765
4. Okamoto N, Hori S, Akazawa C, Hayashi Y, Sigemoto R, Mizuno N, Nakanishi S (1994) Molecular characterization of a new metabotropic glutamate receptor mGluR7 coupled to inhibitory cyclic AMP signal transduction. J Biol Chem 269:1231–1236
5. Desai MA, Conn PJ (1991) Excitatory effects of ACPD receptor activation in the hippocampus are mediated by direct effects on pyramidal cells and blockade of synaptic inhibition. J Neurophysiol (Bethesda) 66:40–52
6. Manzoni O, Prezeau L, Sladeczek F, Bockaert J (1992) Trans-ACPD inhibits cAMP formation via a pertussis toxin-sensitive G-protein. Eur J Pharmacol 225:357–358
7. Henzi V, MacDermott AB (1991) Characteristics and function of Ca^{2+}- and inositol 1,4,5-trisphosphate-releasable stores of Ca^{2+} in neurons. Neuroscience 46:251–273
8. Wakamori M, Hidaka H, Akaike N (1993) Hyperpolarizing muscarinic responses of freshly dissociated rat hippocampal CA1 neurones. J Physiol (Camb) 463:585–604
9. Dolphin AC, Scott RH (1987) Calcium channel currents and their inhibition by (−)-baclofen in rat sensory neurones: modulation by guanine nucleotides. J Physiol (Camb) 386:1–17
10. Tamaoki T, Nomoto H, Takahashi I, Kato Y, Morimoto M, Tomita F (1986) Staurosporine, a potent inhibitor of phospholipid/Ca^{2+}-dependent protein kinase, Biochem Biophys Res Commun 135:397–401
11. Ebihara S, Akaike N (1993) Potassium currents operated by thyrotropin-releasing hormone in dissociated CA1 pyramidal neurones of rat hippocampus. J Physiol (Camb) 472:689–710
12. Houamed KM, Kuijper JL, Gilbert TL, Haldeman BA, O'Hara PJ, Mulvihill ER, Almers W, Hagen FS (1991) Cloning, expression, and gene structure of a G protein-coupled glutamate receptor from rat brain. Science 252:1318–1321
13. Nicoletti F, Wroblewski JT, Fadda E, Costa E (1988) Pertussis toxin inhibits signal transduction at a specific metabotropic glutamate receptor in primary cultures of cerebellar granule cells. Neuropharmacology 27:551–556
14. Watson S, Abbott A (1991) K^+ channels. Trends Pharmacol Sci (TIPS receptor nomenclature supplement) 31–33
15. Abe T, Sugihara H, Nawa H, Shigemoto R, Mizuno N, Nakanishi S (1992) Molecular characterization of a novel metabotropic glutamate receptor mGluR5 coupled to inositol phosphate/Ca^{2+} signal transduction. J Biol Chem 267:13361–13368
16. Adamson P, Hajimohammadeza I, Brammer MJ, Campbell IC, Meldrum BS (1990) Presynaptic glutamate/quisqualate receptors: effects on synaptosomal free calcium concentrations. J Neurochem 55:1850–1854
17. Sladeczek F, Récasens M, Bockaert J (1988) A new mechanism for glutamate receptor action: phosphoinositide hydrolysis. Trends Neurosci 11:545–549
18. Sugiyama H, Ito I, Hirono C (1987) A new type of glutamate receptor linked to inositol phospholipid metabolism. Nature 325:531–533
19. Tanabe S, Ito I, Sugiyama H (1991) Possible heterogeneity of metabotropic glutamate receptors induced in *Xenopus* oocytes by rat brain mRNA. Neurosci Res 10:71–77

20. Hayashi Y, Tanabe Y, Aramori I, Masu M, Shimamoto K, Ohfune Y, Nakanishi S (1992) Agonist analysis of 2-(caboxyclycloprópyl) glycine isomers for cloned metabotropic glutamate receptor subtypes expressed in Chinese hamster ovary cells. Br J Pharmacol 107:539–543
21. Eaton SA, Jane DE, Jones PL, Porter RHP, Pook PCK, Sunter DC, Udvarhelyi PM, Roberts PJ, Salt TE, Watkins JC (1993) Competitive antagonism at metabotropic glutamate receptors by (S)-4-carboxyphenylglycine and (RS)-α-metyle-4-carboxyphenylglycine. Eur J Pharmacol 244:195–197
22. Guérineau NC, Gähwiler BH, Gerber U (1994) Reduction of resting K^+ current by metabotropic glutamate and muscarinic receptors in rat CA3 cells: mediation by G-protein. J Physiol (Camb) 474:27–33
23. Staub C, Vraresic I, Knöpfel T (1992) Responses to metabotropic glutamate receptor activation in cerebellar Purkinje cells: induction of an inward current. Eur J Neurosci 4:832–839
24. Shirasaki T, Harata N, Akaike N (1994) Metabotropic glutamate response in acutely dissociated hippocampal CA1 pyramidal neurones of the rat. J Physiol (Camb) 475:439–453
25. Rainnie DG, Holmes KH, Gallagher PS (1994) Activation of postsynaptic metabotropic glutamate receptors by *trans*-ACPD hyperpolarizes neurons of the basolateral amygdala. J Neurosci 14:7208–7220

Long-Term Potentiation and Drugs

MASAMICHI SATOH[1] and TAKEHIKO MAEDA[2]

Summary. We investigated the modulation of tetanus-induced long-term potentiation (LTP) in guinea pig hippocampal slices brought about by bath-applied drugs and by neurotransmitter-related agents, using an extracellular recording technique. The magnitude of LTP at mossy fiber–CA3 synapses was augmented by the muscarinic receptor agonist carbachol and by endogenous acetylcholine through muscarinic M_1 receptors, but the magnitude was attenuated through M_2 receptors. Somatostatin augmented mossy fiber LTP; the augmentation was reversed by the muscarinic antagonist scopolamine, suggesting that cholinergic neurons were involved. The magnitude of LTP at mossy fiber–CA3 synapses was augmented by bifemelane and depressed by the opioid antagonist naloxone, but was not affected by the *N*-methyl-D-aspartate (NMDA) antagonist D-APV (D-2-amino-5-phosphonovalerate). These drugs had no effect on the LTP at fimbrial fiber–CA3 or Schaffer collateral–CA1 synapses, except for a blockade evoked by D-APV. We also demonstrated, using the slice-patch technique, that the mossy fiber–CA3 LTP was independent of postsynaptic Ca^{2+} increase, membrane depolarization, and the activity of G proteins, and that, in contrast, all three of these factors were required for the expression of fimbrial–CA3 LTP. These observations suggest that the LTP at mossy fiber synapses is induced principally by presynaptic mechanisms, and that it is susceptible to various drugs and neurotransmitter-related agents. It is expected that there may be some parallelism between the effects of drugs on mossy fiber–CA3 LTP in vitro and the effects of these drugs on memory function in vivo.

Key words. Long-term potentiation—Hippocampus—Slice—Extracellular recording—Whole-cell recording—Glutamate—Acetylcholine—Somatostatin—Nootropics—Opioid

Introduction

Long-term potentiation (LTP) of synaptic transmission is a ubiquitous form of use-dependent synaptic plasticity that is present at most excitatory synapses in the central nervous system (CNS). The study of LTP in structures involved in learning and

---

Departments of [1]Molecular Pharmacology and [2]Pharmacology, Faculty of Pharmaceutical Sciences, Kyoto University, Sakyo-ku, Kyoto 606-01, Japan memory, such as the hippocampus and amygdala, has led to the idea that LTP may underlie certain mnemonic functions in the CNS. LTP was first described in the hippocampus [1]. Histochemical and immunohistochemical studies have shown that the hippocampus consists of various types of neurons: cholinergic neurons from the septum [2], noradrenergic neurons from the locus ceruleus [3], serotonergic neurons from the raphe nuclei [4], and somatostatin-containing neurons. If LTP is involved in learning and memory, then it is conceivable that both these processes could be subject to modulation by neurotransmitters and hormones. Indeed, much behavioral evidence indicates that such agents modulate memory processes [5–8]. Thus, the elucidation of hippocampal LTP could contribute greatly to the development of agents for the treatment of dementia.

A major advance in the investigation of hippocampal LTP was the finding that LTP in Schaffer collateral–CA1 pyramidal cell synapses (CA1 LTP) was dependent on the activation the *N*-methyl-D-aspartate (NMDA) subclass of glutamate receptors on postsynaptic neurons [9]. This finding accelerated research into the mechanisms responsible for the production of LTP at the synapses. However, LTP at mossy fiber–CA3 synapses (mossy fiber LTP) was found to be independent of NMDA receptor activation [10], and the mechanism underlying its induction and expression is still unclear.

This review focuses on two points: the modulation of mossy fiber LTP by neurotransmitter-related agents and the mechanisms whereby the LTP is produced.

Drugs that Modulate LTP Production

Tetanus-Induced LTP Production in Hippocampal Slices

Transverse hippocampal slices (500 μm thick) were prepared from male guinea pigs, as described previously [11]. The field potential was recorded in the CA1 or CA3 regions by an extracellular recording technique. Electrical stimulation, given with a bipolar stimulating electrode, was applied to the stratum radiatum of the CA2 region to stimulate Schaffer collaterals or commissural/associational fibers (C/A); to the granule cell layer of the dentate gyrus to stimulate mossy fibers; and to the fimbria to stimulate fimbrial fibers (Fig. 1a). When the respective fibers were stimulated at low frequency (test stimulation), a biphasic field potential (population spike, PS; Fig. 1b) was recorded in the pyramidal cell layer, and field excitatory postsynaptic potential (fEPSP; Fig. 1c) was recorded in the dendritic region. A high-frequency stimulation (tetanic stimulation, TS) of the respective fibers increased the amplitude of the PS and fEPSP. The average magnitude was stable from 12 to 62 min after the TS (Fig. 1d). These phenomena were regarded as the LTP. In some experiments, the area between the zero line and the curve of the PS amplitude from 12 to 62 min after TS was calculated in each slice as another value reflecting the magnitude of the LTP (hatched region in the plot of the LTP time-course shown in Fig. 1d).

Role of Cholinergic Neurons in Mossy Fiber LTP

Anatomical studies of the hippocampus have identified the projections of cholinergic fibers that originate from the medial septum nuclei and terminate in the pyramidal cells and granule cells of the hippocampal formation [2]. Disturbances in brain cho-

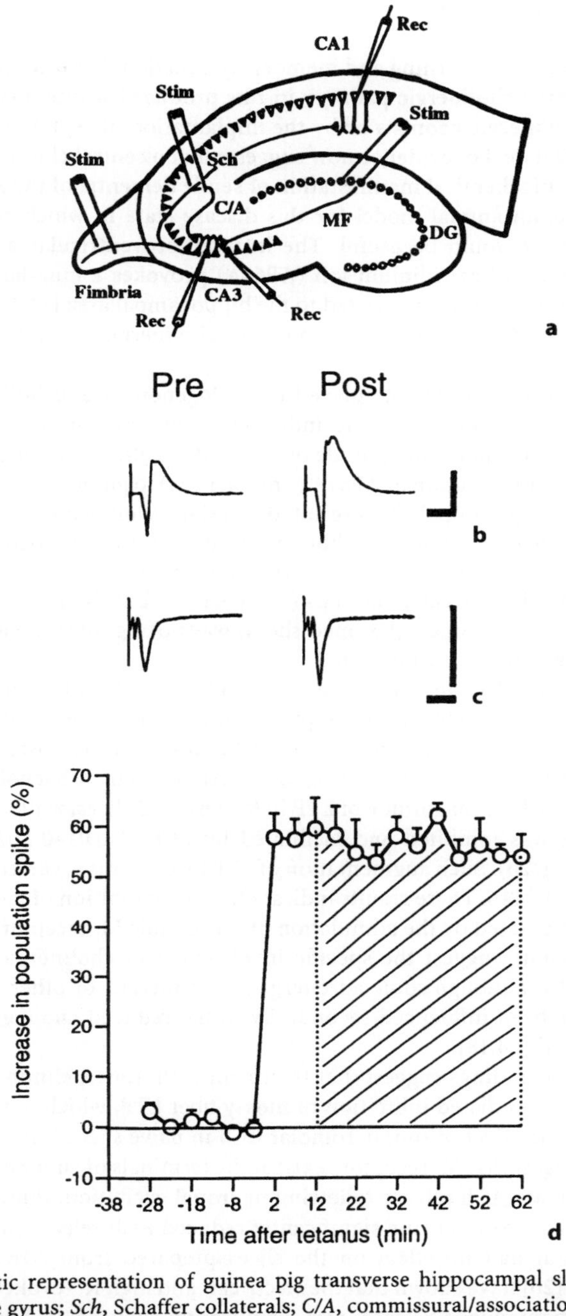

FIG. 1. **a** Schematic representation of guinea pig transverse hippocampal slice. *MF*, Mossy fibers; *DG*, dentate gyrus; *Sch*, Schaffer collaterals; *C/A*, commissural/associational fibers; *Stim*, stimulation electrode; *Rec*, recording electrode. **b** Long-term potentiation (LTP) of the population spike (PS) in CA3 region. **c** LTP of the field excitatory postsynaptic potential (fEPSP) in the CA3 region. The responses were recorded in no-drug slices. *Left* and *right traces* are the responses before (*Pre*) and 27 min after (*Post*) tetanus stimulation (TS), respectively, of the same slice. Calibration: vertical, 1 mV; horizontal, 10 ms. **d** Time-course of LTP in no-drug slices. Ordinate, percentage increase in the amplitude of the PS; abscissa, time after TS in minutes. *Points* and *vertical bars* represent the mean and SE, respectively. The area between the zero line and each curve of TS was calculated (*hatched region*)

linergic neurons cause learning and memory dysfunction. For example, lesions of the septohippocampal cholinergic pathway in rats produce learning deficits that can be reversed by cholinergic agonists or by the implantation of septal neurons [12]. Furthermore, a great deal of evidence now suggests that a central cholinergic deficit may be the primary biochemical manifestation of senile dementia of the Alzheimer's type [13]. Therefore, an animal model for this disease state in which reduced function has been induced would be useful. The intracerebroventricular administration of ethylcholine mustard aziridinium ion (AF64A) provokes a long-lasting decrease in several cholinergic markers restricted to the hippocampal area [14,15], and this agent has been used to develop an animal model of cholinergic disruption and cognitive deficiency [16].

To clarify the functional role played by endogenous acetylcholine (ACh) in the mossy fiber LTP, we evaluated the influence of the reduced cholinergic function provoked by AF64A on the magnitude of LTP and the effects of muscarinic agents on LTP [17]. The administration of AF64A (5 nmol, i.c.v.) to guinea pigs 1 week before the preparation of hippocampal slices resulted in a significant decrease in the magnitude of LTP. The bath application of a cholinesterase inhibitor, physostigmine, at 0.1 μM and 10 μM, attenuated and augmented, respectively, the magnitude of LTP in slices prepared from vehicle-treated animals (naive slices), but the LTP in slices prepared from AF64A-treated animals (lesioned slices) was not significantly affected by physostigmine at any concentration tested.

The induction of LTP in naive slices was inhibited and facilitated, respectively, by the muscarinic M_1 antagonist pirenzepine (1 μM), and by the M_2 antagonist AF-DX 116 (1 μM), alone, while the induction of LTP in lesioned slices was not significantly changed by either of these agents (Fig. 2a). Bath-applied carbachol (CCh), at 0.01–10 μM, augmented the magnitude of LTP in the lesioned slices, while induction of LTP in naive slices was inhibited and facilitated by CCh at 0.01–0.1 μM and 1–10 μM, respectively (Fig. 2b). Such augmentation of LTP by CCh was reversed by pirenzepine, but not by AF-DX 116. These results indicate that the inhibition of mossy fiber LTP by AF64A is compensated by the stimulation of muscarinic M_1 receptors with CCh, and, further, the results suggest the specific involvement of cholinergic neurons in the action of AF64A, although some adrenergic, serotonergic, or other peptidergic neurons may have been influenced secondarily to the reduced cholinergic function, as previously noted [14,18].

The foregoing findings suggest that treatment with AF64A eliminated the muscarinic M_2 receptor-mediated inhibition of mossy fiber LTP, which inhibition appeared to result from the effect of submicromolar CCh in naive slices (Fig. 2b). Several lines of evidence suggest that M_2 receptors exist at the terminals of cholinergic neurons and negatively regulate ACh release following neuronal excitation. Potter and Nitta [19] have shown that oxotremorine significantly reduced ACh release in normal hippocampal slices but had no effect on the slices prepared from AF64A-injected rats. Leventer et al. [20] have shown that the electrical- and high K^+-evoked release of ACh, but not the spontaneous release, was significantly reduced in hippocampal slices from AF64A-treated rats. It has also been reported that hippocampal muscarinic autoreceptors are pharmacologically an M_2 receptor subtype [21]. Therefore, the augmenting effect of submicromolar CCh on mossy fiber LTP in AF64A-lesioned slices may be attributed to the AF64A-induced decrease in the number of presynaptic M_2 autoreceptors, which autoreceptors negatively regulate ACh release, and then to

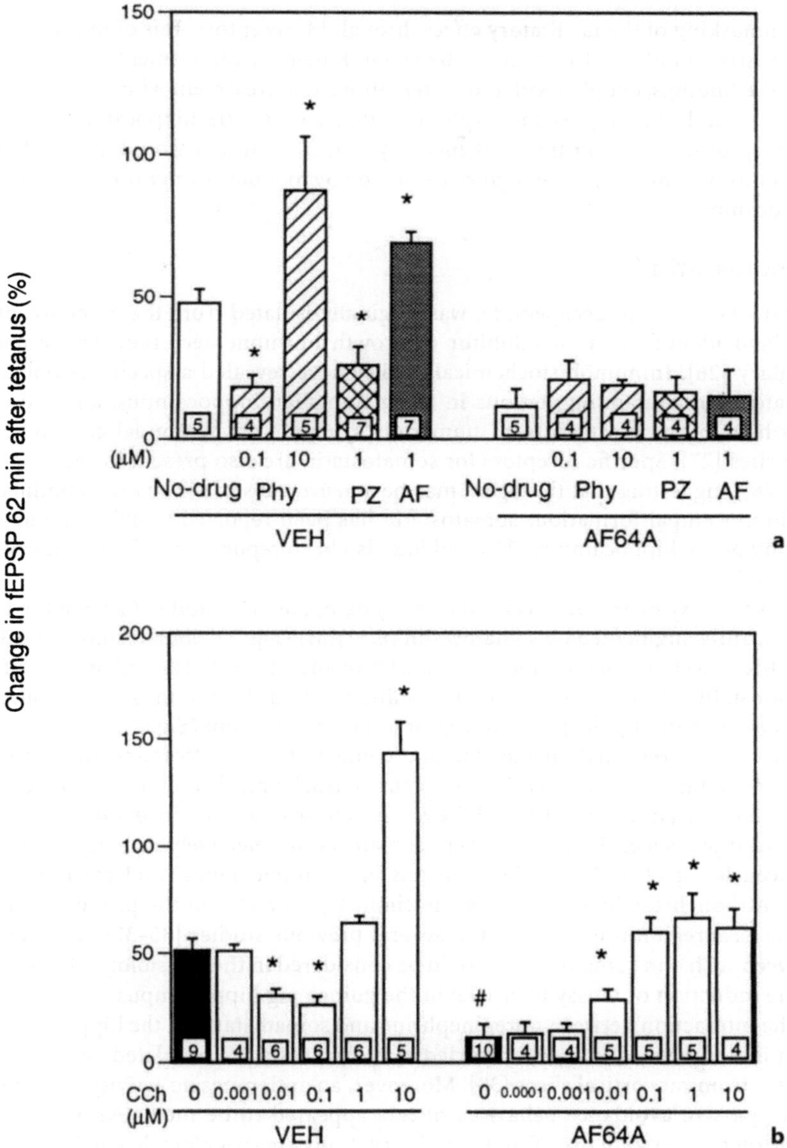

Fig. 2. a Effects of muscarinic agents on LTP in naive and lesioned slices. Ordinate, percent change in *fEPSP* 62 min after tetanus; *column* and *bars*, means ± SE. *Phy*, Physostigmine; *PZ*, pirenzepine; *AF*, AF-DX 116. *, $P < .05$ vs. no-drug naive slices (Mann-Whitney *U*-test). **b** Concentration-dependent effects of carbachol (*CCh*) on mossy fiber LTP in naive and lesioned slices. *, $P < .05$ *vs* no-drug slices from respective treated animals; #, $P < .05$ *vs* no-drug naive slices. Numbers of tested slices are shown in the columns. *VEH*, vehicle

the unmasking of the facilitatory effect through M_1 receptors. It is of interest that the administration of AF64A produced behavioral memory impairment [22,23].

These findings, coupled with our observations that treatment with AF64A inhibited mossy fiber LTP, support the hypothesis that LTP in the hippocampus is an elementary process of learning and memory [24,25]. Thus, it would appear that endogenous ACh may regulate cognitive function by modulating neuronal activity in the hippocampus.

Somatostatin

Somatostatin, a tetradecapeptide, was originally isolated from the ovine hypothalamus and identified as an inhibitor of growth hormone secretion by the anterior pituitary [26]. Immunohistochemical studies have revealed a specific population of somatostatin-containing neurons in the mammalian hippocampus, and these neurons have been shown to form numerous synapses on pyramidal cell bodies and dendrites [27]. Specific receptors for somatostatin are also present in the hippocampus [28], suggesting that this agent may be a neurotransmitter or neuromodulator in the hippocampal formation. Somatostatin has been reported to affect the neuronal activity of the hippocampus [29], and has also been reported to affect memory function [30].

In the mossy fiber–CA3 pyramidal cell system, the magnitude of LTP of the PS was significantly augmented by somatostatin (0.32 μM) (Fig. 3). Somatostatin, at 0.32 μM, failed to affect the magnitude of CA1 LTP of the PS [31]. The enhancing effect of somatostatin on mossy fiber LTP was inhibited both by the muscarinic antagonist scopolamine and by the β-adrenoceptor antagonist timolol (Fig. 3).

Cholinergic terminals make synaptic contacts on somatostatin-immunoreactive cells in the rat hippocampus [32]. A recent neurochemical study revealed that somatostatin presynaptically enhanced the K^+-evoked release of endogenous ACh from rat hippocampal slices [33]. Moreover, somatostatin selectively enhanced the ACh-induced firing of CA1 and CA3 neurons in rat hippocampal and cortical neurons [34]. Although the involvement of the cholinergic system in the production of LTP in the CA3 region was suggested in several previous studies [35–37], an interaction between ACh and somatostatin should be considered in the physiological modulation of the induction of mossy fiber LTP in the guinea pig hippocampus.

The interaction between norepinephrine and somatostatin in the hippocampus has been investigated, and it was reported that somatostatin stimulated norepinephrine release from rat cortical slices [38]. Moreover, an antiamnesiac activity of somatostatin in passive avoidance behaviors of rats appeared to be mediated by the central noradrenergic system [39]. From our observations, we speculate that an interaction of norepinephrine and somatostatin in the hippocampus may play a physiological role in the development of mossy fiber LTP. On the other hand, LTP at the CA1 pyramidal neurons following tetanic stimulation of Schaffer collaterals was not affected by somatostatin at a concentration (0.32 μM) that was effective in mossy fiber LTP. This finding suggests that somatostatin is involved not in the development of CA1 LTP, but in the modulation of mossy fiber LTP.

Different Drug Susceptibilities of LTP in the CA3 Region

It is known that the pyramidal neurons of the CA3 region receive at least three different inputs, the mossy fibers, C/A fibers, and fimbrial fibers [40]. In each input

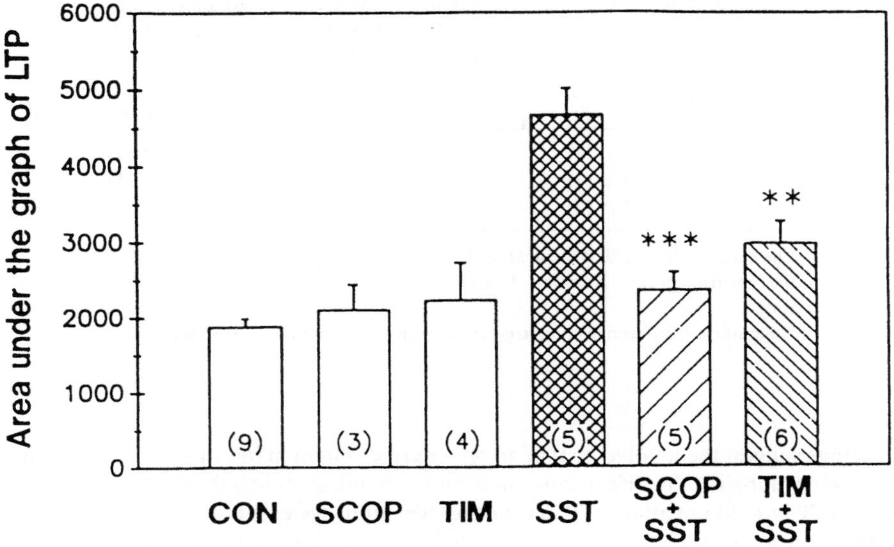

FIG. 3. Effects of simultaneous application of somatostatin with scopolamine (1 μ*M*) or timolol (1 μ*M*) on LTP of PS in mossy fiber–CA3 system. **, *P* < .01; ***, *P* < .001, vs. somatostatin alone (Student's *t*-test). *CON*, Control slices; *SCOP*, 1 μ*M* scopolamine; TIM, 1 μ*M* timolol; SST, 0.32 μ*M* somatostatin. *SCO* + *SST* and *TIM* + *SST* show simultaneous application of scopolamine and timolol, respectively, with somatostatin

system, LTP has been demonstrated in slice preparations [10,41,42]. Harris and Cotman [10] demonstrated that D-2-amino-5-phosphonovalerate (D-APV) inhibited LTP in the C/A–CA3 system (C/A LTP) but not mossy fiber LTP. These observations prompted us to investigate the effects of several different types of drugs, the NMDA antagonist, D-APV, naloxone (an opioid antagonist), and bifemelane, on LTP in the three input systems to the CA3 region.

The effects of these three drugs on LTP in the CA3 region are shown in Table 1. D-APV blocked C/A LTP, but not mossy fiber LTP, in the hippocampal slices of the guinea pig, as reported previously [10]. Moreover, the drug also inhibited LTP in the fimbria–CA3 system (fimbrial LTP), which LTP is suggested to be mediated by NMDA receptors in a manner similar to that operating with C/A LTP. In the CA3 region, mossy fibers terminate in the stratum lucidum, while C/A and fimbrial fibers terminate in the stratum radiatum and stratum oriens [43]. Monaghan et al. [44] reported that the density of NMDA receptors was low and that of kainate receptors was high in the stratum lucidum of the CA3 region, but that the density of quisqualate and NMDA receptors was high in the stratum radiatum and stratum oriens.

In contrast with the effect of D-APV, we found that naloxone blocked mossy fiber LTP, a finding consistent with that previously reported by Martin [45]. Furthermore, it has been demonstrated that naloxone does not affect LTP in the other two input systems to the CA3 region. It has been revealed, by an immunohistochemical technique, that hippocampal mossy fibers are dense with dynorphin A, an opioid peptide [46]. Further, Satoh et al. [47] showed that dynorphin A augmented LTP in the mossy fiber–CA3 system and that such an augmenting effect was blocked by naloxone. These

TABLE 1. Effects of D-2-amino-5-phosphonovalerate (D-APV) (10 μ*M*), naloxone (1 μ*M*), and bifemelane (1 μ*M*) on long-term potentiation (LTP) in the mossy fiber–, commisural/associational fiber–, and fimbria–CA3 systems

Drug	Mossy fiber	Commissural/associational fiber	Fimbria
D-APV	0.9 (3)	0.1* (3)	0.1* (3)
Naloxone	0.3* (4)	1.0 (3)	1.2 (3)
Bifemelane	3.1* (4)	1.1 (3)	0.9 (3)

The magnitude of LTP is the area under the graph in Fig. 3 compared to that for the no-drug group. Numbers in parentheses indicate the number of slices tested.

*, Significant difference compared with no-drug group in each system.

findings suggest the involvement of an opioidergic system in mossy fiber LTP, but not in C/A or fimbrial LTP. Bifemelane, on the other hand, significantly augmented mossy fiber LTP but did not affect LTP in the other two input systems to the CA3 region. CA1 LTP was previously shown not to be affected by this drug [48]. These findings indicate that C/A LTP and fimbrial LTP are similar to CA1 LTP.

Mechanism of Induction of Mossy Fiber LTP

No Requirement for Postsynaptic $[Ca^{2+}]_i$ and Depolarization

The mechanisms underlying the induction of LTP have been most intensively studied in the CA1 region of the hippocampus. In this region, high-frequency stimulation of Schaffer collateral/commissural input activates postsynaptic NMDA receptors and causes a marked membrane depolarization that removes the voltage-dependent Mg^{2+} block of NMDA receptor channels, through which Ca^{2+} enters postsynaptic neurons and triggers LTP. Hippocampal CA3 pyramidal neurons receive three excitatory inputs, of which LTP at the mossy fiber–CA3 synapse has distinct characteristics. Unlike CA1 LTP or LTP in the other two inputs to CA3 pyramidal neurons, the induction of mossy fiber LTP does not require the activation of NMDA receptor [10,49,50]. Furthermore, mossy fiber LTP does not show associative properties [51]. Therefore, the mechanisms underlying the induction of mossy fiber LTP may be distinct from those responsible for CA1 LTP.

The patch-clamp recording technique has been employed with brain slice preparations [52]; with this technique, control of membrane potential and of the intracellular milieu can be achieved, which has contributed greatly to advances in investigations of the detailed mechanisms underlying synaptic transmission and its modification. In our laboratory, we have employed the whole-cell patch-clamp recording technique with CA3 pyramidal cells in thin hippocampal slices to elucidate the mechanisms underlying LTP in the CA3 region [53].

Tight-seal whole-cell recordings were carried out in hippocampal slice preparations, essentially according to the methods described by Edwards et al. [52]. The hippocampi of 20- to 27-day-old rats were cut into 200-μ*M*-thick slices. CA3 pyramidal neurons were viewed with an upright microscope equipped with Nomarski

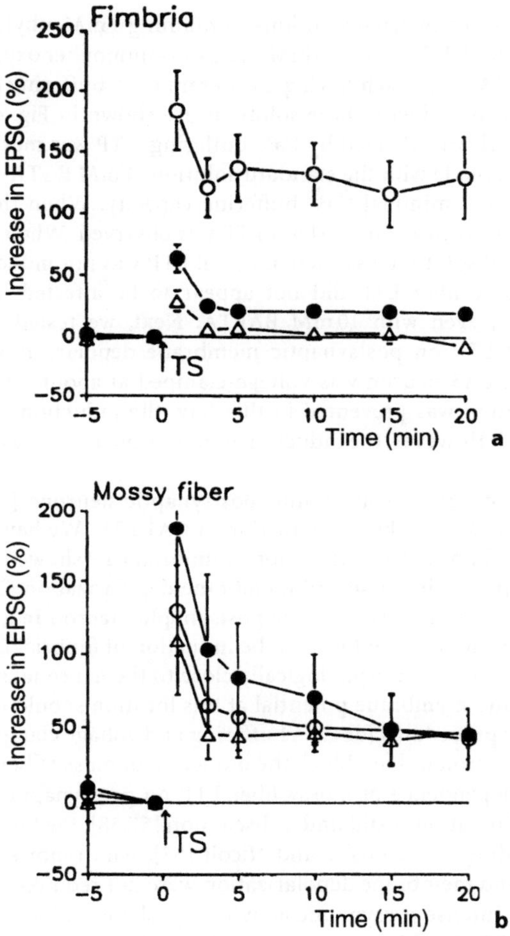

FIG. 4. a Effect of Ca^{2+} chelators applied intracellularly on the induction of fimbrial LTP. *Open circles*, internal solution containing 1 m*M* ethyleneglycoltetraacetic acid (EGTA) ($n = 6$); *closed circles*, internal solution containing 10 m*M* EGTA ($n = 7$); *open triangles*, internal solution containing 10 m*M* 1,2-bis(2-aminophenoxy)ethane-N,N,N′,N′-tetraacetic acid (BAPTA) ($n = 5$). **b** Mossy fiber LTP was not affected by Ca^{2+} buffering. In all cases, TS of mossy fibers was delivered in the presence of 25–50 µ*M* D-2-amino-5-phosphonovalerate (D-APV) ($n = 5$ for all three groups). *, $P < .05$; #, $P < .01$, *vs* cells treated with 1 m*M* EGTA in the experiments on mossy fiber LTP

optics. Stimulating pulses were delivered through a bipolar electrode positioned in the dentate granule cell layer or fimbria, and the excitatory postsynaptic currents (EPSC) of the CA3 pyramidal neurons were recorded in a patch-clamp mode (Fig. 4).

As it has been suggested that Ca^{2+} plays an important role in the generation of LTP [54], we examined the effect of the buffering of intracellular Ca^{2+} on the induction of LTP. We used three internal solutions, containing 1 m*M* ethyleneglycoltetraacetic acid (EGTA), 10 m*M* EGTA, or 10 m*M* 1,2-bis(2-aminophenoxy)ethane-N,N,N′,N′-tetraacetic acid (BAPTA), each having a different Ca^{2+}-buffering capacity. The time-courses of LTP recorded with these solutions are shown in Fig. 4. The induction of fimbrial LTP was clearly affected by Ca^{2+} buffering. LTP was markedly elicited when the recording was made with the standard solution (1 m*M* EGTA-containing internal solution), which had minimal Ca^{2+}-buffering capacity. When the concentration of EGTA was raised to 10 m*M*, only a slight LTP was observed. When the 10 m*M* BAPTA-containing internal solution was used, fimbrial LTP was not induced. In contrast, the induction of mossy fiber LTP did not appear to be affected by the buffering of intracellular Ca^{2+}, even with 10 m*M* BAPTA. Next, we tested the dependence of the induction of LTP on postsynaptic membrane depolarization (Fig. 5). TS was delivered while a CA3 neuron was voltage-clamped at about −80 mV. When membrane depolarization was prevented in this way, the induction of fimbrial LTP was largely attenuated. However, the induction of mossy fiber LTP was not affected by this procedure.

The injection of Ca^{2+} chelators into postsynaptic neurons [55] or postsynaptic hyperpolarization [56] blocks the induction of CA1 LTP. We have shown that this is also the case with fimbrial LTP. Induction of fimbrial LTP showed a clear dependence on postsynaptic $[Ca^{2+}]_i$ increase and membrane depolarization. Therefore, it is conceivable that Ca^{2+}, which flows into the postsynaptic neuron in a voltage-dependent manner, may play an important role in the induction of fimbrial LTP. Because mossy fiber synapses are located morphologically close to the neuronal soma [51], manipulation of $[Ca^{2+}]_i$ and membrane potential at this location should be equally effective. Nevertheless, our procedures for Ca^{2+} buffering and voltage clamping in the postsynaptic pyramidal neuron did not block the induction of mossy fiber LTP. Although the question of the dependence of mossy fiber LTP on postsynaptic $[Ca^{2+}]_i$ increase or membrane depolarization is still under discussion [57,58], the foregoing results are in line with the findings of Zalutsky and Nicoll [57], who reported that postsynaptic $[Ca^{2+}]_i$ increase and membrane depolarization were not required for the induction of mossy fiber LTP. Intense presynaptic activity is probably sufficient for the induction of mossy fiber LTP. In this context, using the outside-out patch-clamp recording technique to detect excitatory amino acid release from hippocampal slices, we recently found increased release of excitatory amino acids, but not change in postsynaptic responsiveness, during mossy fiber LTP [59].

Involvement of Postsynaptic G-proteins

Guanosine 5-triphosphate- (GTP-) binding proteins (G proteins) are involved in many cellular signal transduction processes. The involvement of G proteins in LTP has been suggested by the observation that pretreatment of animals with pertussis toxin suppresses both mossy fiber LTP [60] and CA1 LTP [61]. However, these pretreatment experiments have not clarified the location of G proteins (pre- or postsynaptic, or in the interneurons) responsible for LTP. Goh and Pennefather [61] have reported that the postsynaptic injection of guanosine 5′-*O*-(3-thiotriphosphate) (GTPγS) does not occlude CA1 LTP, thus indicating that postsynaptic G proteins are not responsible for CA1 LTP. However, there is no direct evidence to show whether postsynaptic G proteins are involved in LTP in the CA3 region.

We studied the possible involvement of postsynaptic G proteins in LTP in rat

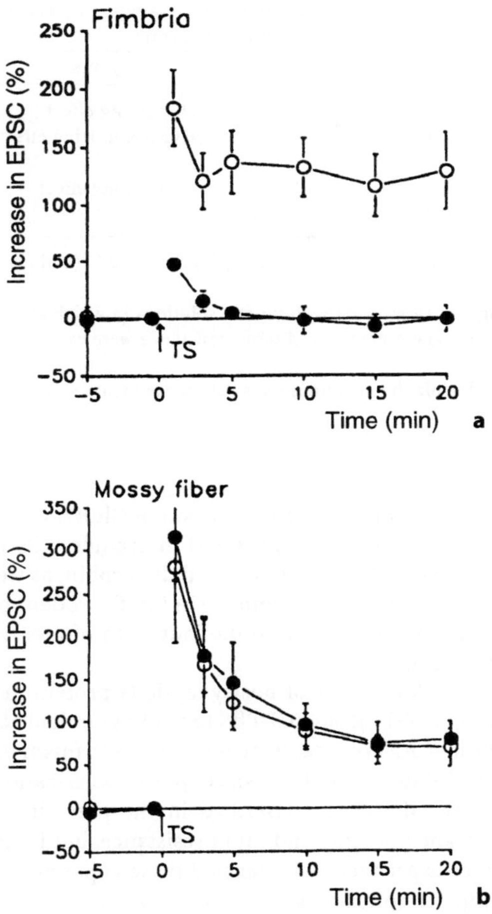

Fig. 5. **a** Dependence of induction of fimbrial LTP on postsynaptic membrane depolarization. *Open circles*, delivery of TS under current clamp; *closed circles*, delivery of TS under voltage clamp at −80 mV. The internal solution contained 1 m*M* EGTA. **b** The induction of mossy fiber LTP was not dependent on postsynaptic membrane depolarization. The internal solution containined 10 m*M* EGTA. *EPSC*, Excitatory postsynaptic currents

hippocampal slices, using a whole-cell recording technique with a recording pipette containing GTPγS and guanosine 5′-[β-thio] diphosphate (GDPβS) [62]. The effects of GTP analogs on the passive membrane properties of CA3 pyramidal neurons and on both fimbrial and mossy fiber LTP are summarized in Table 2. We found that the induction of mossy fiber LTP was not affected by GDPβS or by GTPγS applied to a CA3 pyramidal cell, supporting the postulate that the mossy fiber LTP utilizes a mechanism that is confined to the presynaptic element.

Because in vivo treatment with pertussis toxin abolishes mossy fiber LTP [60], the location of the pertussis toxin-sensitive G proteins responsible for this LTP is probably in the mossy fiber terminals. Candidates for the receptor involved in the generation of the mossy fiber LTP via G proteins are the metabotropic glutamate receptor (mGluR) and opioid receptors. Among the mGluR subtype family, both mGluR1,

Table 2. Effects of guanosine triphosphate (GTP) analogs on passive membrane properties and LTP in two input systems to CA3 pyramidal neurons

Property	GTPγS	GDPβS
Input resistance[a]	Decreasing effect	
Resting membrane Potential[a]	Hyperpolarizing effect	
Mossy fiber LTP		
Fimbrial LTP	Attenuating effect	Attenuating effect
Fimbrial LTP[b] (forced depolarization during TS)		

GTPγS, Guanosine 5′-*O*-(3-thiotriphosphate); GDPβS, guanosine 5′-(B-thio) diphosphate; TS, tetanus stimulation.

[a] Whole-cell recording was carried out with pipette solutions in which each of the GTP analogs was dissolved. Resting membrane potential and input resistance were measured 15–30 min after whole-cell access.

[b] Tetanus was delivered while the membrane was depolarized from −80 to −20 mV by voltage command pulses.

which is linked to phospholipase C activation, and mGluR2, which is linked to adenylate cyclase inhibition, are strongly expressed in granule cells in the dentate gyrus [63,64]. Immunoreactivities for enkephalin and dynorphin, as well as opioid-binding sites, are localized in the stratum lucidum of CA3 in the rodent hippocampus [65]. A contribution of the μ opioid receptor to the induction of mossy fiber LTP has been also demonstrated [45,49,66,67].

In contrast to the lack of effect of postsynaptic G-protein modulation on mossy fiber LTP, the loading of GTPγS and GDPβS to CA3 pyramidal cells alters the magnitude of LTP at fimbria–CA3 synapses. One possible mechanism underlying the effects of GTP analogs is the inhibition of the postsynaptic depolarization that is required for the induction of fimbrial LTP, as described in the previous chapter. GDPβS did not affect the membrane potential and input resistance, but it did reduce the magnitude of LTP. Another experiment has clarified these apparently inconsistent results. When K^+ in the pipette solution was replaced by Cs^+ to block some K^+ channel opening, and TS was delivered while the membrane was depolarized to −20 mV by a voltage command, the inhibition evoked by GTPγS and GDPβS was almost entirely reversed. These observations indicate that the inhibition evoked by GDPβS is mainly the result of insufficient depolarization during TS, and that GDPβS specifically reduces the depolarization induced by TS without affecxting the resting K^+ conductance.

We found, by extracellular recording [68], that the magnitude of fimbrial LTP was reduced by atropine when the intensity of TS was rather weak (50 Hz, 1 s), but that this magnitude was not altered when a strong TS was delivered (three trains of 100 Hz, 1 s). In contrast, a β-adrenergic antagonist, timolol, attenuated the LTP after mild TS. No apparent changes were produced with either mild or strong TS by an mGluR antagonist, DL-2-amino-3-phosphonopropionate. Although the locations of the relevant receptors have not yet been identified, these observations suggest that fimbrial LTP is modulated by an endogenous neurotransmitter through G-protein-linked receptors. A comparison of the neurotransmitter modulation of LTP at fimbrial and mossy fiber synapses should be helpful in increasing our understanding of the way in which individual afferent plasticity regulates the excitability of postsynaptic CA3 pyramidal neurons.

Conclusion

The production of mossy fiber LTP in hippocampal slices is modulated by neurotransmitters and nootropics. However, the susceptibility of mossy fiber LTP to some drugs differs from that of LTP in other systems in hippocampal slices. Furthermore, the production of LTP in the fimbria–CA3 system, but not in the mossy fiber–CA3 system, was blocked by postsynaptic manipulations during TS. These results indicate that the mechanisms whereby LTP is produced and modulated in mossy fiber–CA3 synapses differ from those of LTP in other synapses in hippocampal slices. The detailed mechanisms underlying LTP production in the hippocampus have yet to be revealed.

References

1. Bliss TVP, Lømo T (1973) Long-lasting potentiation of synaptic transmission in the dentate area of the anaesthetized rabbit following stimulation of the perforant path. J Physiol 232:331–356
2. Frotscher M, Leranth C (1985) Cholinergic innervation of the rat hippocampus as revealed by choline acetyltransferase immunocytochemistry: a combined light and electron microscopic study. J Comp Neurol 239:237–246
3. Loy R, Koziell DA, Lindsey JD, Moore RY (1980) Noradrenergic innervation of the adult rat hippocampal formation. J Comp Neurol 189:699–710
4. Freund TF, Gulyas AI, Acsady L, Gorcs T, Toth K (1990) Serotonergic control of the hippocampus via local inhibitory interneurons. Proc Natl Acad Sci USA 87:8501–8505
5. Bartus RT, Dean RL, Bernard B, Lippa AS (1982) The cholinergic hypothesis of geriatric memory dysfunction. Science 217:408–418
6. Crow TJ, Wendlandt S (1976) Impaired acquisition of a passive avoidance response after lesions induced in the locus ceruleus by 6-OH dopamine. Nature 259:42–44
7. Barnes MJ, Costall B, Coughlan J, Domeney AM, Gerrad AP, Keely EM, Natlor JR, Onaivi SE, Tomkins MD, Tyers BM (1990) The effects of ondansetron, 5-HT_3 receptor antagonist, on cognition in rodents and primates. Pharmacol Biochem Behav 35:955–962
8. Vecsei L, Bollok I, Telegdy G (1983) Intracerebroventricular somatostatin attenuates electroconvulsive shock-induced amnesia. Peptides 4:293–295
9. Collingridge GL, Kehl SJ, McLennan H (1983) The antagonism of amino acid-induced excitations of rat hippocampal CA1 neurons in vitro. J Physiol (Camb) 334:19–31
10. Harris EW, Cotman CW (1986) Long-term potentiation of guinea-pig mossy fiber responses is not blocked by *N*-methyl-D-aspartate antagonists. Neurosci Lett 70:132–137
11. Iwama T, Ishihara K, Satoh M, Takagi H (1986) Different effects of dynorphin A on guinea-pig hippocampal CA3 pyramidal cells with various degrees of paired-pulse facilitation. Neurosci Lett 63:190–194
12. Low WC, Lewis PR, Bunch ST, Dunnett SB, Thomas SR, Iversen SD, Bjorklund A, Stenevi U (1982) Function recovery following neural transplantation of embryonic septal nuclei in adult rats with septohippocampal lesions. Nature 300:260–262
13. Coyle JT, Price DL, Delong M (1983) Alzheimer's disease: a disorder of cortical cholinergic innervation. Science 219:1184–1190
14. Hörtnagl H, Potter PE, Hanin I (1987) Effect of cholinergic deficit induced by ethylcholine aziridinium on serotonergic parameters in rat brain. Neuroscience 22:203–213
15. Potter PE, Tedford C, Kindel G, Hanin I (1989) Inhibition of high affinity transport attenuates both cholinergic and non-cholinergic effect of AF64A. Brain Res 487:238–244

16. Fisher A, Hanin I (1986) Potential animal models for senile dementia of Alzheimer's type with particular emphasis on AF64A-induced cholinotoxity. Annu Rev Pharmacol Toxicol 26:161–181
17. Maeda T, Kanko S, Satoh M (1994) Roles of endogenous cholinergic neurons in the induction of long-term potentiation at hippocampal mossy fiber synapses. Neurosci Res 20:71–78
18. Potter PE, Tedford C, Kindel G, Hanin I (1989) Inhibition of high affinity transport attenuates both cholinergic and non-cholinergic effect of AF64A. Brain Res 487:238–244
19. Potter PE, Nitta S (1993) Alteration in modulation of acetylcholine release following lesion of hippocampal cholinergic neurons with the neurotoxin AF64A. Neuropharmacology 32:519–526
20. Leventer S, McKeag D, Clancy M, Wulfert E, Hanin I (1985) Intracerebroventricular administration of ethylcholine mustard aziridinium ion (AF64A) reduces release of acetylcholine from rat hippocampal slices. Neuropharmacology 24:453–459
21. Richards MH (1990) Rat hippocampal autoreceptors are similar to the M_2 (cardiac) subtype: comparison with hippocampal M_1, atrial M_2 and ileal M_3 receptors. Br J Pharmacol 99:753–761
22. Fisher A, Hanin I (1980) Choline analogs as potential tools in developing selective animal methods of central cholinergic hypofunction. Life Sci 27:1615–1634
23. Yamazaki N, Kato K, Kurihara E, Nagaoka A (1991) Cholinergic drugs reverse AF64A-induced impairment of passive learning in rats. Psychopharmacology 103:215–222
24. Berger WT, Rinaldi CP, Weisz JD, Thompson RF (1983) Single-unit analysis of different hippocampal cell types during classical condition of rabbit nictating membrane response. J Neurophysiol (Bethesda) 50:1197–1219
25. Berger WT (1984) Long-term potentiation of hippocampal synaptic transmission affects rate of behavioral learning. Science 224:627–630
26. Brazeau P, Vale E, Burgus R, Ling N, Butcher M, Rivier J, Guillemin R (1973) Hypotheramic polypeptide that inhibits the secretion of immunoreactive pituitary growth hormone. Science 179:77–79
27. Kohler C, Chan-Palay V (1982) Somatostatin-like immunoreactive neurons in the hippocampus: an immunocytochemical study in the rat. Neurosci Lett 34:259–264
28. Wood PL, Cheney DL, Costa E (1979) Modulation of the turnover rate of hippocampal acetylcholine by neuropetides: possible site of action of α-melanocyte-stimulating hormone, adrenocorticotrophic hormone and somatostatin. J Pharmacol Exp Ther 209:97–103
29. Scharfman HE, Scwartzkroin PA (1989) Selective depression of GABA-mediated IPSPs by somatostatin in area CA1 of rabbit hippocampal slices. Brain Res 493:205–211
30. Matsuoka N, Maeda N, Yamaguchi I, Satoh M (1994) Role of somatostatin in the augmentation of hippocampal long-term potentiation by FR121196, a putative cognitive enhancer. Eur J Neuropharmacol 241:27–34
31. Matsuoka N, Kaneko S, Satoh M (1991) Somatostatin augments long-term potentiation of the mossy fiber–CA3 system in guinea-pig hippocampal slices. Brain Res 553:188–194
32. Lerenth C, Frotscher M (1987) Cholinergic innervation of hippocampal GAD- and somatostatin-immunoreactive commissural neurons. J Comp Neurol 261:33–47
33. Araujo DM, Lapchak PA, Collier B, Quirion R (1990) Evidence that somatostatin enhances endogenous acetylcholine release in the rat hippocampus. J Neurochem 55:1546–1555
34. Mancillas JR, Siggins GR, Bloom FE (1986) Somatostatin selectively enhances acetylcholine-induced excitations in rat hippocampus and cortex. Proc Natl Acad Sci USA 83:7518–7521
35. Satoh M, Ishihara K, Katsuki H (1989) A pharmacological profile of LTP in CA3 region of guinea-pig hippocampus in vitro. Biomed Res 10:125–129
36. Tanaka Y, Sakurai M, Hayashi S (1989) Effect of scopolamine and HP 029, a cholinesterase inhibitor, on long-term potentiation in hippocampal slices of the guinea pig. Neurosci Lett 98:197–183

37. Maeda T, Kaneko S, Satoh M (1994) Bidirectional modulation of long-term potentiation by carbachol via M_1 and M_2 muscarinic receptors in guinea pig hippocampal mossy fiber–CA3 synapses. Brain Res 619:324–330
38. Tsujimoto A, Tanaka S (1981) Stimulatory effect of somatostatin on norepinephrine release from rat brain cortex slices. Life Sci 28:903–910
39. Vecsei L, Bollok I, Telegdy G (1984) Phenoxybenzamine antagonizes somatostatin-induced anti-amnesia in rats. Eur J Pharmacol 99:325–328
40. Brodal A (1981) The olfactory pathways. The amygdala. The hippocampus. The "limbic system". In: Brodal A (ed) Neurological anatomy in relation to clinical medicine, 3rd edn. Oxford, New York, pp 640–697
41. Yamamoto C, Chujo T (1978) Long-term potentiation in thin hippocampal sections studied by intracellular and extracellular recordings. Exp Neurol 58:242–250
42. Yamamoto C, Sawada S (1981) Important factors in induction of long-term potentiation in thin hippocampal sections. Exp Neurol 74:122–130
43. Raisman G, Cowan WM, Powel TPS (1965) The extrinsic afferent, commissural and association fibers of the hippocampus. Brain 88:963–996
44. Monaghan DT, Yao D, Cotman CW (1985) L-[^{3}H] Glutamate binds to kainate-, NMDA- and AMPA-sensitive binding sites: an autoradiographic analysis. Brain Res 340:378–383
45. Martin MR (1983) Naloxone and long-term potentiation of hippocampal CA3 field potentials in vitro. Neuropeptides 4:45–50
46. Khachaturian H, Watson SJ, Lewis ME, Coy D, Goldstein A, Akill H (1982) Dynorphin immunocytochemistry in the rat central nervous system. Peptides 3:941–954
47. Satoh M, Ishihara K, Katsuki H, Sugimura M (1989) Opioidergic modulation of long-term potentiatioin in the guinea-pig hippocampus in vitro. Adv Biosci 75:193–196
48. Satoh M, Ishihara K, Katsuki H (1988) Different susceptibilities of long-term potentiation in CA3 and CA1 regions of guinea pig hippocampal slices to nootropic drugs. Neurosci Lett 93:236–241
49. Ishihara K, Katsuki H, Sugimura M, Kaneko S, Satoh M (1990) Different drug-susceptibilities of long-term potentiation in three input systems to the CA3 region of the guinea pig hippocampus in vitro. Neuropharmacology 29:487–492
50. Williams S, Johnston D (1988) Muscarinic depression of long-term potentiation in CA3 hippocampal neurons. Science 242:84–87
51. Brown TH, Johnston D (1983) Voltage-clamp analysis of mossy fiber synaptic input to neurons. J Neurophysiol (Bethesda) 50:487–507
52. Edwards FA, Konnerth A, Sakmann B, Takahashi T (1989) A thin slice preparation for patch clamp recordings from neurons of the mammalian central nervous system. Pflügers Arch 414:600–612
53. Katsuki H, Kaneko S, Tajima A, Satoh M (1991) Separate mechanisms of long-term potentiation in two input systems to CA3 pyramidal neurons of rat hippocampal slices as revealed by the whole-cell patch-clamp technique. Neurosci Res 12:393–402
54. Bliss TVP, Collingridge GL (1993) A synaptic model of memory: long-term potentiation in the hippocampus. Nature 361:31–39
55. Lynch G, Larson J, Kelso S, Barrionuevo G, Schotter F (1983) Intracellular injection of hippocampal long-term potentiation. Nature 305:719–721
56. Malinow R, Miller JP (1986) Postsynaptic hyperpolarization during conditioning reversibly blocks induction of long-term potentiation. Nature 320:529–530
57. Zalutsky RA, Nicoll RA (1990) Comparison of two forms of long-term potentiation in single hippocampal neurons. Science 248:1619–1624
58. Jaffe D, Johnston D (1990) Induction of long-term potentiation at hippocampal mossy fiber synapses follows a Hebbian rule. J Neurophysiol (Bethesda) 64:948–960
59. Maeda T, Mizukami K, Kaneko S, Akaike A, Satoh M (1994) Increased excitatory amino acid release detected with outside-out patch electrode during mossy fiber LTP in rat hippocampal slices. Neurosci Res Suppl 19:S32
60. Ito I, Okada D, Sugiyama H (1988) Pertussis toxin suppresses long-term potentiation of hippocampal mossy fiber synapses. Neurosci Lett 90:181–185

61. Goh JW, Pennefather PS (1989) A pertussis toxin-sensitive G protein in hippocampal long-term potentiation. Science 244:980–983
62. Katsuki H, Kaneko S, Satoh M (1992) Involvement of postsynaptic G proteins in hippocampal long-term potentiation. Brain Res 581:108–114
63. Masu M, Tanabe Y, Tsuchida K, Shigemoto R, Nakanishi S (1991) Sequence and expression of a metabotropic glutamate receptor. Nature 349:760–765
64. Tanabe Y, Masu M, Ishii T, Shigemoto R, Nakanishi S (1992) A family of metabotropic glutamate receptors. Neuron 8:169–179
65. McLean S, Rothman RB, Jacobson AE, Rice KC, Herkenham M (1987) Distribution of opiate receptor subtypes and enkephalin and dynorphin immunoreactivity in the hippocampus of squirrel, guinea pig, rat and hamster. J Comp Neurol 255:497–510
66. Derrick BE, Rodriguez SB, Lieberman DN, Martinez JL Jr (1992) Mu opioid receptors are associated with the induction of hippocampal mossy fiber long-term potentiation. J Pharmacol Exp Ther 263:725–733
67. Satoh M, Ishihara K, Katsuki H (1989) Pharmacological profile of LTP in CA3 region in guinea-pig hippocampus in vitro. Biomed Res 10(suppl 2):125–129
68. Katsuki H, Saitoh H, Satoh M (1992) The involvement of muscarinic, β-adrenergic and metabotropic glutamate receptors in long-term potentiation in the fimbria–CA3 pathway of the hippocampus. Neurosci Lett 142:249–252

Analysis of Distributions of Nitric Oxide Synthase mRNAs in the Normal Rat Brain by In Situ Hybridization Histochemistry

Toshiro Kumanishi[1], Rong Rong Han[1], Kazuo Kobayashi[1], Shu Jing Zhang[1], Tomio Ichikawa[1], Yoko Tsuchiya[2], Toshiaki Nagafuji[3], Toru Matsui[4], Tsutomu Ogura[5], and Hiroyasu Esumi[5]

Summary. Distributions of neuronal nitric oxide synthase (n-NOS) and inducible nitric oxide synthase (i-NOS) mRNAs in the normal adult rat brain were investigated by in situ hybridization histochemistry. On examination of n-NOS mRNA, signals were detected in various regions of the cerebrum, the brainstem, and the cerebellum in their constituent neurons. The quantitative analysis revealed that the numbers of signals per cell were variable, ranging from 5–10 to more than 200, depending on the region. Brain regions without distinct n-NOS signals were also seen. This marked difference in the amount of n-NOS transcripts seemed to indicate differences of contribution of NO in physiological function among neurons of different regions. However, as seen in the cerebral cortex, the caudate/putamen, the hippocampal gyrus, and the dentate gyrus, although most neurons showed no or weak signals, a small number of sporadically scattered neurons showed intense signals, indicating functional heterogeneity of neurons even in the same region. Nonneuronal elements including glial cells and endothelial cells revealed no distinct signals for n-NOS mRNA. On examination of i-NOS mRNA, no distinct signals were detected in the brain parenchyma, the choroid plexus, the meninges, or the blood vessels, suggesting that if i-NOS mRNA was expressed in some cells in the normal rat brain, it might be only in a small amount. These findings might provide a basis for studies on regulation of NOS expression at the transcriptional and posttranscriptional levels in various physiological and pathological states, including brain ischemia. Advantages of the use of pyronin-methylgreen in counterstaining and in determination of the optimal conditions of in situ hybridization procedures are also described.

Key words. Neuronal nitric oxide synthase—Inducible nitric oxide synthase—NOS—Nitric oxide—In situ hybridization—mRNA—Pyronin-methylgreen

---

[1]Department of Molecular Neuropathology, Brain Research Institute, Niigata University, 1-757 Asahimachi-dori, Niigata 951, Japan

[2]Drug Metabolism and Pharmacokinetics Research Laboratory, Chugai Pharmaceutical Co., Ltd., 3-41-8 Takada, Toshima-ku, Tokyo 171, Japan

[3]Fuji Gotemba Research Laboratories, Chugai Pharmaceutical Co., Ltd., 135 Komakado, 1 Gotemba, Shizuoka 412, Japan

[4]Department of Neurosurgery, Saitama Medical Center/School, 1981 Kamoda, Kawagoe, Saitama 350, Japan

[5]Investigative Treatment Division, National Cancer Center Research Institute, East, 6-5-1 Kashiwanoha, Kashiwa, Chiba 277, Japan

Introduction

The family of nitric oxide synthases (NOS; EC 1.14.23) consists of at least three distinct isozymes: neuronal (n), endothelial (e), and inducible (i) NOSs, derived from different genes [1,2]. These NOSs catalyze the synthesis of NO, a gaseous molecule, from L-arginine when they are activated by the binding of calmodulin and some other molecules in a Ca^{2+}-dependent [neuronal NOS (n-NOS) and endothelial NOS (e-NOS)] or Ca^{2+}-independent [inducible NOS (i-NOS)] manner [1,2]. In various cells and tissues, NO is produced in response to various stimuli and shows a variety of biological functions [1,3]. In the nervous system, NO has been suggested to be involved in neurotransmission [3,4], long-term potentiation [5,6], long-term depression [7,8], and also in neurotoxicity [3,9,10]. Although a number of studies have already been conducted concerning the mechanism of these phenomena, a number of unsolved problems remain, including those of regulation of NOS expression in physiological and pathological conditions.

In this study, we performed in situ hybridization histochemistry to investigate the cellular localization of n-NOS and i-NOS mRNA in the normal adult rat brain, which might provide a basis for studies on regulation of NOS expression at the transcriptional and posttranscriptional levels in various physiological and pathological states including brain ischemia.

Experimental Procedures

Probes

n-NOS cDNA, n-NOS oligonucleotide, and i-NOS cDNA were used as probes for in situ hybridization. n-NOS oligonucleotide and i-NOS cDNA were rat sequences, while n-NOS cDNA was a mouse sequence that showed a high homology (93%) to the corresponding region of rat n-NOS cDNA. The approximate locations of these probes along the mRNA sequences are shown schematically in Fig. 1. The n-NOS cDNA probe, 560 bp in size, corresponded to a part within the proximal one-third of the

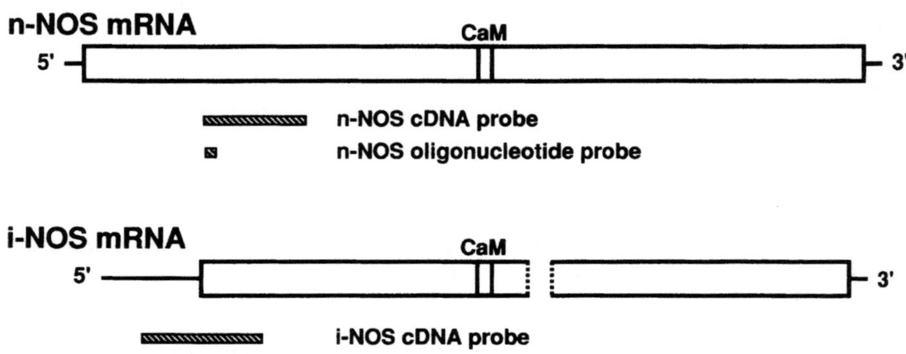

FIG. 1. Schematic of approximate locations of neuronal nitric oxide synthase (*n-NOS*) cDNA, n-NOS oligonucleotide, and inducible nitric oxide synthase (*i-NOS*) cDNA probes along the mRNA sequences. Locations of probes are shown by *thin striped bars* along the mRNAs in which the predicted calmodulin (*CaM*) binding site is indicated. *Thin line*, 5′- or 3′-untranslated region; *open bar*, coding region

coding region of the n-NOS mRNA [11]. When the first nucleotide for the initial methionine was numbered as +1, the nucleotide positions (np) of the n-NOS cDNA probe were +664 to +1223, which showed a relatively low homology (about 40%) to that of the corresponding region of i-NOS mRNA [12]. The n-NOS cDNA probe was ^{35}S-labeled using a random primed DNA labeling kit (Boehringer Mannheim, Indianapolis, IN, USA) to a specific activity of about 5×10^8 cpm/μg. The n-NOS oligonucleotide probe (antisense, 5′-CATCGAGGTCTCTGTCCACCTG GATTCCTGTGTCTTTTCACTTCTG-3′) corresponded to a part of the n-NOS cDNA probe, np +677–+721, which showed a 37% homology to that of the corresponding region of i-NOS mRNA. The oligonucleotide probe was 3′-end labeled with [^{35}S]dATP (DuPont/NEN, Boston, MA, USA) using a terminal deoxyribonucleotidyl transferase (Takara, Tokyo, Japan) to a specific activity of about 5×10^8 cpm/μg.

These n-NOS oligonucleotide and cDNA probes were expected to hybridize not only the ordinary form of n-NOS mRNA, but also its variant forms, which were produced by alternative splicing either in the 5′-untranslated region [1,13] or in the middle of n-NOS mRNA [14], both locations being outside the target sequences for the probes.

The i-NOS cDNA probe, 658 bp in size, corresponded to a region of i-NOS mRNA [12] spanning the 3′-end of the 5′-untranslated region and the 5′-end of the coding region, np −286–+372, of which the part for the coding region (+1–+372) showed a 36% homology to that of the corresponding region of n-NOS mRNA. The i-NOS cDNA probe was ^{35}S-labeled to a specific activity of about 5×10^8 cpm/μg by the same method as described for n-NOS cDNA.

Procedure for In Situ Hybridization

Brain tissues were removed from anesthetized adult Wistar and Sprague-Dawley (SD) rats and immediately embedded and frozen in O.C.T. compound (Miles, Kankakee, IL, USA) using cold *n*-hexan in dry ice-acetone. Sagittal and coronal sections, 12 μm thick, were made on a cryostat, attached to 0.01% poly-L-lysine-coated glass slides, and fixed in 4% paraformaldehyde/0.1 *M* sodium phosphate buffered saline (PBS) (pH 7.2) for 30 min. The sections were washed with PBS containing 2 mg/ml of glycine for 20 min and acetylated in 0.25% acetate/0.1 *M* Tris-HCl (pH 8.0) for 10 min, followed by washing with 2 × SSC (1 × SSC = 0.15 *M* NaCl, 0.015 *M* sodium citrate) for 5 min three times.

The sections were then prehybridized for 3 h at 42°C in a prehybridization solution containing 50% formamide, 5 × SSC, 50 m*M* sodium phosphate (pH 6.5), 2% Sarkosyl, 1× Denhardt's solution (0.02% Ficoll, 0.02% polyvinyl-pyrrolidone, 0.02% bovine serum albumin), 250 μg/ml heat-denatured salmon sperm DNA, and 10 m*M* β-mercaptoethanol, and hybridized for 16 h at 42°C in a solution containing 4 vol of prehybridization solution, 1 vol of 50% dextran sulfate, and heat-denatured ^{35}S-labeled cDNA probe (2×10^4 cpm/μl) or oligonucleotide probe (5×10^3 cpm/μl). The tissue sections were then washed three times in 0.1 × SSC/0.1% Sarkosyl at 42°C for 40 min, dehydrated, and autoradiographed using NTB2 nuclear track emulsion (Kodak, Rochester, NY, USA) for 2 months at 4°C.

After development, the sections were counterstained with pyronin-methylgreen, as described later, and examined with a light microscope. In some experiments, the hybridized sections were exposed to Hyperfilm-β max (Amersham, Little Chalfont, UK) for 8 days before autoradiography with NTB2 film emulsion.

In the control competition experiment for n-NOS mRNA, the sections were hybridized with a mixture of the radiolabeled oligonucleotide probe and a 25-fold excess of unlabeled probe.

Quantitative Analysis of Silver Grains

With a light microscope, silver grains were quantified in the representative cells of each neuronal cell type using autoradiographed sections counterstained with pyronin-methylgreen.

Pyronin-Methylgreen Staining

For counterstaining, autoradiographed sections were thoroughly dried, stained with a pyronin-methylgreen stain solution (Muto Pure Chemicals, Tokyo, Japan) for 1 h, washed in distilled water for 1–2 s, dried rapidly using an air-dryer, immersed in xylene, and mounted in balsam.

Because intracellular RNAs were stained with pyronin [15], the same stain solution was also used to determine whether intracellular RNAs in sections were fixed adequately for subsequent successive procedures of in situ hybridization, and to determine whether the reagents for these procedures were free from RNase-like activity. In the former case, frozen sections were fixed with 4% paraformaldehyde/PBS for various time periods, washed, processed with the same procedures of in situ hybridization, excluding autoradiography, and then washed, dehydrated, stained with pyronin-methylgreen as described, and examined for the degree of intracellular pyronin staining. In the latter case, frozen sections were fixed with 4% paraformaldehyde/PBS for 30 min, washed, incubated with a given reagent solution to be tested for 1 h at 42°C, and then washed, dehydrated, stained with pyronin-methylgreen, and examined for the degree of intracellular pyronin staining.

Distribution of n-NOS mRNA

In the in situ hybridization analysis of the rat brain, signals for n-NOS mRNA were detected in neurons in some regions, but not distinctly in others. The brain regions with distinct n-NOS signals are listed in Table 1. In most of these regions, signals appeared similar in most neurons in a given region, but the signal intensities in neurons were variable depending on the region. When estimated by the quantitative analysis, the numbers of silver grains per cell ranged from 5–10 to more than 200, depending on the region (Table 1). Signals were found to be intense in neurons of the pedunculopontine tegmental nucleus, followed by those in the supraoptic nucleus, the subthalamic nucleus, and some of the pontine tegmental nuclei. Signals were moderate in neurons of the accessory olfactory bulb, the amygdala, the paraventricular hypothalamic nuclei, the superior and the inferior colliculi, and in stellate/basket cells in the cerebellum. Signals were relatively weak in neurons of the olfactory bulb, the piriform cortex, the olfactory tubercle, the hippocampal gyrus, the dentate gyrus, and the granule cell layer of the cerebellum. Figure 2 shows n-NOS signals in the representative neurons.

Signals were not distinct in neurons of the cerebral cortex, the caudate/putamen, and the thalami in the cerebrum, in Purkinje cells and neurons of the dentate nucleus in the cerebellum, and in neurons of the pontine base. Among these regions, however,

Table 1. Brain regions with distinct neuronal nitric oxide synthase (n-NOS) signals

Brain region	Signal intensity
Olfactory bulb	+
Accessory olfactory bulb	++
Piriform cortex	+
Olfactory tubercle	+
Amygdala	++
Hippocampus (CA1–CA4)	+
Dentate gyrus	+
Hypothalamus	
Supraoptic nucleus	+++
Subthalamic nucleus	+++
Others	++
Superior colliculus	++
Inferior colliculus	++
Pedunclopontine tegmental nucleus	++++
Pontine tegmentum	
Supragenual nucleus	+++
Lateral and medial parabrachial nucleus	+++
Cerebellum	
Granule cell layer	+
Stellate/basket cells	++

The numbers of silver grains per cell were roughly estimated in representative cells of each neuronal cell type. The number of background silver grains is less than 3 when estimated in small cells. Symbols: +, 5–10 grains/cell; ++, 10–20 grains/cell; +++, 20–100 (usually 30–50) grains/cell; ++++, 100 to >200 grains/cell.

the cerebral cortex and the caudate/putamen revealed a small number of sporadically scattered neurons with intense signals, showing a marked contrast to the background field without distinct signals (Fig. 2f). The hippocampal gyrus and the dentate gyrus also revealed a few neurons with intense signals around the pyramidal cell layer and the granule cell layer, respectively, in both of which most of the neurons showed weak signals (Fig. 2e). Glial cells in the callosal body and the cerebellar white matter revealed no distinct n-NOS signals, although the callosal body contained a few cells with intense signals that might possibly be neuronal cells of heterotopic nature (data not shown). Endothelial cells in the large and small blood vessels also showed no distinct n-NOS signals (data not shown). All these findings were consistently obtained with either an n-NOS cDNA probe or an oligonucleotide probe in both Wistar and SD rats.

Control sections hybridized with a ^{35}S-labeled oligonucleotide probe in the presence of a 25-fold excess of unlabeled oligonucleotide probe revealed a remarkable decrease in signals (Fig. 3).

Examination of i-NOS mRNA

The i-NOS cDNA probe revealed no distinct signals in any cellular elements, including glial cells, in the white matter and endothelial cells of blood vessels in the brain parenchyma, the choroid plexus, and the meninges (data not shown).

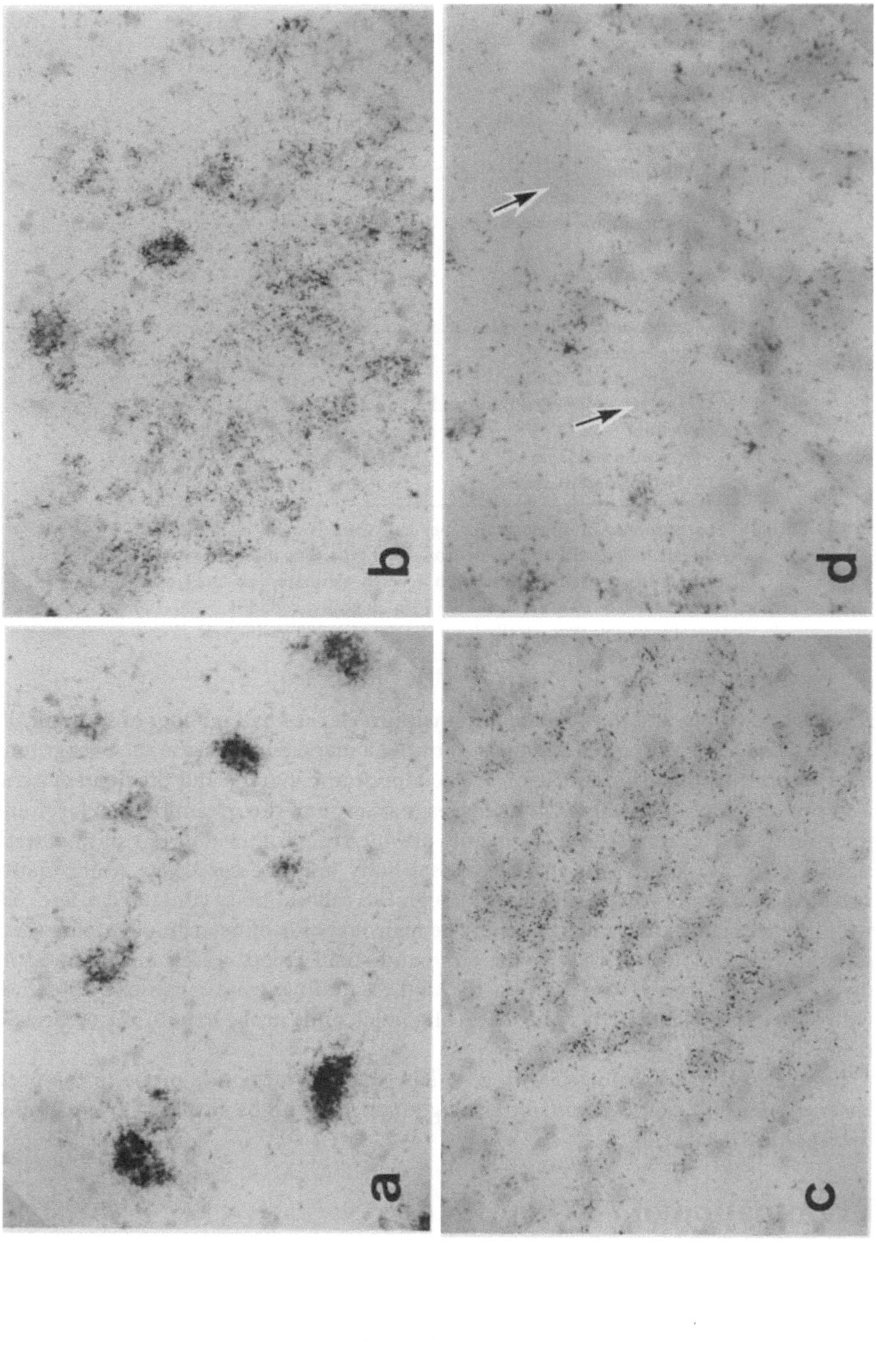
a
b
c
d

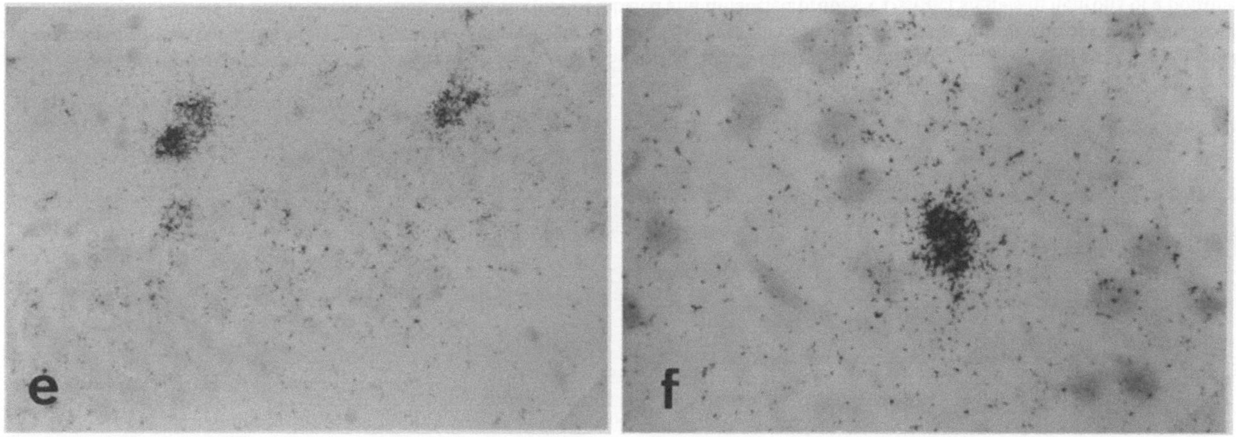

Fig. 2a–f. In situ hybridization of adult rat brain with ^{35}S-labeled n-NOS cDNA probe; **a–g** Counterstained with pyronin-methylgreen. **a** Intense signals in neurons of the pedunculopontine tegmental nucleus. ×300. **b** Intense signals in neurons of the supraoptic nucleus. ×300. **c** Weak signals in neurons of the olfactory bulb. ×300. **d** Moderate signals in stellate/basket cells in the molecular layer (*upper left*), weak signals in neurons of the granule cell layer (*lower right*), and no distinct signals in Purkinje cells (*arrows*) in the cerebellum. ×600. **e** Weak signals in neurons of the granule cell layer (*lower left to upper right*) of the dentate gyrus and intense signals in a few neurons around the granule cell layer. ×300. **f** Intense signals in a neuron in the caudate/putamen in which most neurons show no distinct signals. ×600

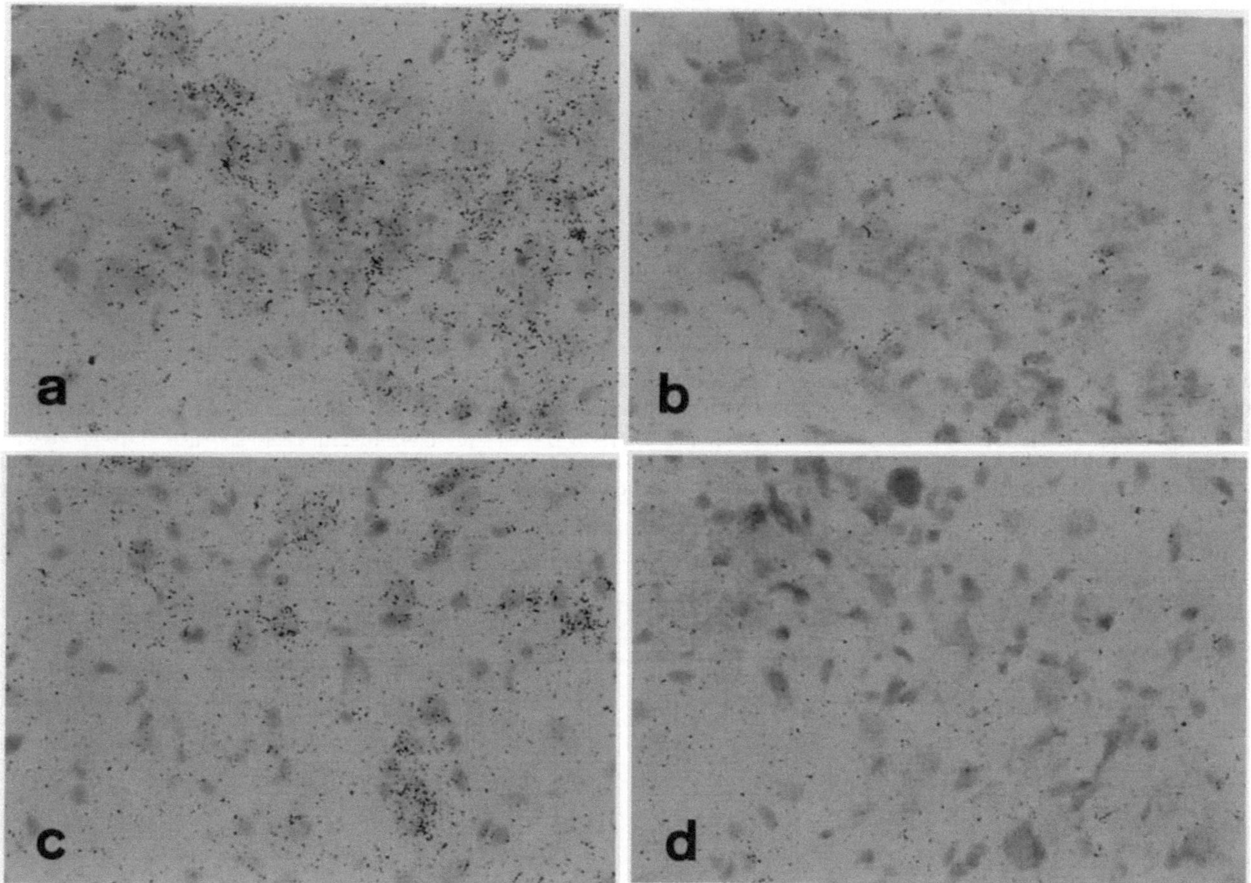

Fig. 3a–d. Control competition experiment. Sections were hybridized with ^{35}S-labeled n-NOS oligonucleotide probe (**a, c**) or with a mixture of labeled and a 25-fold excess of unlabeled probe (**b, d**). Counterstain, pyronin-methylgreen, ×330. **a** Distinct signals in neurons of the supraoptic nucleus with the labeled probe. **b** Control for **a**. No distinct signals in the adjacent section with a mixture of labeled and unlabeled probes. **c** Distinct signals in neurons of a pontine tegmentum nucleus with the labeled probe. **d** Control for **c**. No distinct signals were seen in the adjacent section with a mixture of labeled and unlabeled probes

Pyronin-Methylgreen Staining in In Situ Hybridization

In sections fixed for 30 min with 4% paraformaldehyde and stained with pyronin-methylgreen, the nuclei of cells were stained blue and the cytoplasm reddish-pink. The cytoplasmic reddish-pink staining was especially remarkable in large neurons. This cytoplasmic staining disappeared with treatment of the sections with RNase, indicating that stained materials were mostly RNAs (data not shown).

When this staining was used for counterstaining in in situ hybridization, signals were precisely localized in cellular elements in which the preservation of intracytoplasmic RNAs could also be monitored (see Fig. 3). In addition, the solid film emulsion covering the sections was not stained, allowing a high resolution in microscopic analysis.

This staining was also used to pinpoint the optimal fixation time of the sections and to determine whether reagent solutions for in situ hybridization were free from RNase-like activity. When the sections were fixed for 30 min or more with 4% paraformaldehyde and processed successively with the procedures of in situ hybridization, intense reddish-pink staining was seen in the cytoplasm, especially in those of large neurons. In contrast, sections unfixed, or fixed for less than 10 min, revealed disappearance of the cytoplasmic staining, indicating that RNAs fixed insufficiently were not preserved in sections during subsequent in situ hybridization procedures (data not shown). During the examination of reagent solutions for in situ hybridization, the presence of RNase-like activity indicated by the decrease or disappearing of cytoplasmic reddish-pink staining in sections was seen in some of the commercial reagents, including bovine serum albumin (BSA), which was used in preparation of Denhardt's solution (data not shown).

Discussion

In this study, we performed in situ hybridization histochemistry to investigate the distribution of n-NOS and i-NOS mRNAs in the adult rat brain. On examination of n-NOS mRNA, signals were detected in some regions of the cerebrum, the brainstem, and the cerebellum, as described in Table 1. In these regions, signals were localized in most constituent neurons. However, as shown by the quantitative analysis, the numbers of signals per cell varied with the region. Brain regions without distinct n-NOS signals were also seen. This marked difference in the amount of n-NOS transcripts might indicate differences of contribution of NO in physiological function among neurons of different regions. However, in the cerebral cortex, the caudate-putamen, the hippocampal gyrus, and the dentate gyrus, although most neurons showed no or weak signals, a small number of sporadically scattered neurons showed intense signals, indicating functional heterogeneity of neurons even in the same region. These findings regarding localization of n-NOS mRNA in the normal rat brain were approximately consistent with the previous work of Bredt et al. [11], except for our detection of signals in some additional regions.

An alternatively spliced n-NOS mRNA that was 315 nucleotides and 105 amino acid residues shorter than the ordinary form because of a lack of sequences from 2 exons in the middle of mRNA has been demonstrated in mouse brain [14] and human neuronal cell lines [16]. The amount of this variant form of n-NOS mRNA (n-NOS-2) was estimated to be about 5% of the ordinary n-NOS mRNA (n-NOS-1) by the quantitative polymerase chain reaction (PCR) analysis [14]. The corresponding variant n-NOS mRNA was also detected in the rat brain (T. Ogura et al., unpublished data), suggesting that the presence of these two different forms of n-NOS was common in various animal species.

In a separate experiment, we have performed in situ hybridization to investigate the distribution of these two mRNAs in the mouse brain (data not shown). The oligonucleotide probe for the ordinary n-NOS-1 form (48-mer) corresponded to a part of a conjugated sequence of these 2 exons, and that for the variant n-NOS-2 form (48-mer) corresponded to a sequence which was a conjugation of 24 mer each from the adjacent proximal and distal exons. While the use of the former probe revealed almost the same distribution pattern of signals as the current observation of rat n-NOS mRNA, the use of the latter probe revealed no distinct signals in any brain regions. Thus, although the n-NOS probes used in this study corresponded to the sequences upstream of the region for the alternative splicing and therefore were expected to recognize both the ordinary and the variant forms of n-NOS mRNA, our current findings in the rat brain represented mainly the distribution of the ordinary n-NOS-1 form. The variant n-NOS-2 might be expressed only in a small amount at the single cell level.

More recently, several additional variants of n-NOS mRNA have also been detected. These variants had different 5′-untranslated regions generated by the use of different exons [1,13]. Although the reason for the presence of these variant forms was unknown, it was possible that these were differently expressed depending on the cell type or the state of the intracellular molecular substance. Our study might potentially detect all these variants and might not distinguish one from another because the target sequence for our n-NOS probes was expected to occur commonly in these variants. Thus, further studies using probes specific for individual variant forms might be of particular value for better understanding of the entire profile of n-NOS expression in the brain.

As compared with the use of NTB2 film emulsion in autoradiography, the use of X-ray film was more convenient for rapid evaluation of an approximate distribution of hybridized signals. In this study, we also examined X-ray film images of n-NOS signals in the rat brain and observed considerably more intense signals in the olfactory bulb and the granular cell layer of the cerebellum (data not shown). These findings coincided with the previous work of Bredt et al. [11] in which a mixture of three different oligonucleotide probes was used and the signal density for n-NOS mRNA obtained was estimated by low-power view in a dark field. Actually, however, we detected relatively weak or at most moderate signals in individual cells in these structures in the quantitative analysis, as described. This discrepancy might be interpreted in connection with the high cellularity of small neurons in these structures in which the signal density in the two- or three-unit dimensions might be high despite relatively weak signals in individual cells. This should be taken into consideration in evaluation of the signal intensity in X-ray film images in general.

Although the existence of i-NOS has been suggested in glial cells [17,18] and endothelial cells [1,19] in some in vitro experiments, we failed to detect signals for i-NOS mRNA in glial cells in the cerebral and the cerebellar white matter or in blood vessels in the brain parenchyma, the choroid plexus, and the meninges. This suggested that if i-NOS mRNA was expressed in these cells, it might be only in a small amount in the normal rat brain.

In this study, we used pyronin-methylgreen in the counterstaining of hybridized sections. We were able to localize signals in connection with cell shape and cell type, monitoring the degree of preservation of intracellular RNA. It seemed of additional value that, distinct from other counterstainings, the solid film emulsion covering sections remained unstained, allowing high resolution in microscopic analysis. Moreover, this staining was also used to determine optimal conditions for in situ hybridization procedures and to monitor whether or not the reaction solutions used were free from RNase-like activity. Although more than 90% of intracellular RNA was rRNA and therefore the pyronin-stained materials should mostly be rRNA, we considered the degree of preservation of mRNA to be represented by that of the rRNA estimated by this staining. Unless the rRNA was well preserved, the mRNA might not be expected to be preserved enough for in situ hybridization. Thus, we recommend the use of this simple, rapid, and sensitive pryonin-methylgreen staining for in situ hybridization.

As described in the introduction, NO was considered to be involved in neurotransmission, long-term potentiation, long-term depression, and neurotoxicity in the nervous system. Regarding neurotoxicity, increased NO has been demonstrated in ischemic animal brains [20–23], and this has been considered to play a role in the increase of necrotic volume [24–27]. Immunohistochemically, increases in n-NOS-positive neurons [28] and e-NOS-positive blood vessels [29] have been reported in the ischemic rat brain. However, the mechanism underlying these phenomena remained largely unknown. Some conflicting data were also reported [30]. This study might provide a basis for further studies on the details of these questions.

Acknowledgment. We thank Drs. N. Naito, A. Muto, and M. Sugiyama of Fuji-Gotemba Research Laboratories, Chugai Pharmaceutical Company, Ltd., for their cooperation and helpful suggestions. Moreover, we wish to thank Ms. C. Honma of the Department of Molecular Neuropathology, Brain Research Institute, Niigata University, for her skillful technical assistance. This study was supported in part by a Grand-in Aid from the Ministry of Education, Science and Culture of Japan.

References

1. Nathan C, Xie Q-W (1994) Regulation of biosynthesis of nitric oxide. J Biol Chem 269:13725–13728
2. Schmidt HHHW, Hofmann H, Ogilvie P, Sennefelder H, Weinberg RJ (1994) Biochemistry and regulation of nitric oxide synthase. In: Takagi H, Toda N, Hawkins RD (eds) Nitric oxide: roles in neuronal communication and neurotoxicity. Japan Scientific Societies Press, Tokyo, pp 3–18
3. Dawson TM, Dawson VL, Snyder SH (1992) A novel neuronal messenger molecule in brain: the free radical, nitric oxide. Ann Neurol 32:297–311
4. Bredt DS, Snyder SH (1992) Nitric oxide, a novel neuronal messenger. Neuron 8:3–11
5. Böhme GA, Bon C, Stutzmann J-M, Doble A, Blanchard J-C (1991) Possible involvement of nitric oxide in long-term potentiation. Eur J Pharmacol 199:379–381
6. Schuman EM, Madison DV (1991) A requirement of the intercellular messenger nitric oxide in long-term potentiation. Science 254:1503–1506
7. Shibuki K, Okada D (1991) Endogenous nitric oxide release required for long-term synaptic depression in the cerebellum. Nature 349:326–328

8. Shibuki K (1994) Nitric oxide as a modulator of cerebellar synaptic plasticity. In: Takagi H, Toda N, Hawkins RD (eds) Nitric oxide: roles in neuronal communication and neurotoxicity. Japan Scientific Societies Press, Tokyo, pp 161–169
9. Dawson VL, Dawson TM, London ED, Bredt DS, Snyder SH (1991) Nitric oxide mediates glutamate neurotoxicity in primary cortical cultures. Proc Natl Acad Sci USA 88:6368–6371
10. Dawson VL, Dawson TM, Bartley DA, Uhl GR, Snyder SH (1993) Mechanisms of nitric oxide-mediated neurotoxicity in primary brain cultures. J Neurosci 13:2651–2661
11. Bredt DS, Hwang PM, Glatt CE, Lowenstein C, Reed RR, Snyder SH (1991) Cloned and expressed nitric oxide synthase structurally resembles cytochrome P-450 reductase (GeneBank accession number X59949). Nature 351:714–718
12. Adachi H, Iida S, Oguchi S, Ohshima H, Suzuki H, Nagasaki K, Kawasaki H, Sugimura T, Esumi H (1993) Molecular cloning of a cDNA encoding an inducible calmodulin-dependent nitric-oxide synthase from rat liver and its expression in COS 1 cells. Eur J Biochem 217:37–43
13. Ogura T (1995) Regulation of nitric oxide synthase in the nervous system (in Japanese). Exp Med 13:912–916
14. Ogura T, Yokoyama T, Fujisawa H, Kurashima Y, Esumi H (1993) Structural diversity of neuronal nitric oxide synthase mRNA in the nervous system. Biochem Biophys Res Commun 193:1014–1022
15. Unna PG (1913) The origin of plasma cells (in German). Virchows Arch 214:320–339
16. Fujisawa H, Ogura T, Kurashima Y, Yokoyama T, Yamashita J, Esumi H (1994) Expression of two types of nitric oxide synthase mRNA in human neuroblastoma cell lines. J Neurochem 63:140–145
17. Simmons ML, Murphy S (1992) Induction of nitric oxide synthase in glial cells. J Neurochem 59:897–905
18. Hokari A, Zeniya N, Esumi H (1994) Cloning and functional expression of human inducible nitric oxide synthase (NOS) cDNA from a glioblastoma cell line A-172. J Biochem Tokyo 116:575–581
19. Kilbourn RG, Belloni P (1990) Endothelial cell production of nitrogen oxides in response to interferon gamma in combination with tumor necrosis factor, interleukin-1, or endotoxin. J Natl Cancer Inst 82:772–776
20. Malinski T, Bailey F, Zhang ZG, Chopp M (1993) Nitric oxide measured by a porphyrinic microsensor in rat brain after transient middle cerebral artery occlusion. J Cereb Blood Flow Metab 13:355–358
21. Matsui T, Nagafuji T, Tsutsumi K, Itoh S, Nagata K, Asano T (1994) Direct measure of nitric oxide in rat brain subjected to permanent and temporary ischemia. Abstr Soc Neurosci 20:1479
22. Matsui T (1995) Direct measurement of nitric oxide in the rat brain subjected to occlusion of the middle cerebral artery. In: Ohnishi ST, Ohnishi T (eds) Central nervous system trauma: reseach techniques. CRC series in membrane-linked diseases. CRC Press, Boca Raton, FL, pp 469–482
23. Matsui T, Nagafuji T, Asano T, Kumura E, Yoshimine T, Hayakawa T (1994) Possible participation of nitric oxide in ischemic cerebral injury (in Japanese). Shinkei Kenkyu Shinpo (Adv Neurol Sci) 38:957–966
24. Matsui T, Nagafuji T, Auer RN, Koide T, Tsutsumi K, Asano T (1993) Beneficial effect of nitric oxide synthase inhibitor on reversible and permanent focal cerebral ischemia in rats, using newly devised miniclip. J Cereb Blood Flow Metab 13(suppl):S150
25. Nagafuji T, Matsui T, Koide T, Asano T (1992) Blockade of nitric oxide formation by N^{ω}-nitro-L-arginine mitigates ischemic brain edema and subsequent cerebral infarction in rats. Neurosci Lett 147:159–162
26. Buisson A, Margaill I, Callebert J, Plotkine M, Boulu RG (1993) Mechanisms involved in the neuroprotective activity of a nitric oxide synthase inhibitor during focal cerebral ischemia. J Neurochem 61:690–696
27. Nowicki JP, Duval D, Poignet H, Scatton B (1991) Nitric oxide mediates neuronal death after focal cerebral ischemia in the mouse. Eur J Pharmacol 204:339–340

28. Zhang ZG, Chopp M, Gautam S, Zaloga C, Zhang RL, Schmidt HHHW, Pollock JS, Förstermann U (1994) Upregulation of neuronal nitric oxide synthase and mRNA, and selective sparing of nitric oxide synthase-containing neurons after focal cerebral ischemia in rat. Brain Res 654:85–95
29. Zhang ZG, Chopp M, Zaloga C, Pollock JS, Förstermann U (1993) Cerebral endothelial nitric oxide synthase expression after focal cerebral ischemia in rats. Stroke 24:2016–2022
30. Yamamoto S, Golanov EV, Berger SB, Reis DJ (1992) Inhibition of nitric oxide synthesis increases focal ischemic infarction in rat. J Cereb Blood Flow Metab 12:717–726

Calcium-Dependent Neuronal Cell Death

Takeshi Nakamura[1], Akihiko Ogura[2], and Yoshihisa Kudo[1]

Summary. The mechanism(s) underlying ischemia-induced neuronal death are discussed with special respect to the massive release of glutamate and the excessive accumulation of Ca^{2+} inside the neurons. Energy deprivation caused by ischemia should lead to depolarization of the cell membrane because of the retardation of Na^+/K^+-ATPase. Such depolarization will activate the voltage-gated Ca^{2+} channels and cause an increase in $[Ca^{2+}]_i$. This energy deprivation will impede the Ca^{2+} pump located on the plasma membrane, which is one cause of the elevation of $[Ca^{2+}]_i$. Furthermore, depolarization of the neuronal cell will result in glutamate release, which enhances glutamate release self-regeneratively. Released glutamate should induce $[Ca^{2+}]_i$ elevation through the depolarization-induced disinhibition of the *N*-methyl-D-aspartate (NMDA) receptor. $[Ca^{2+}]_i$ also increases through the reversal of Na^+/Ca^{2+} antiporter and phosphorylation-induced activation of the Ca^{2+} channel. Elevated $[Ca^{2+}]_i$ might result in activation of Ca^{2+}-dependent enzymes, which upregulate self-regeneratively through partial autolysis of protease or self-phosphorylation. In addition to such rapid necrotic cell death, we discuss the mechanisms of delayed neuronal cell death that occur several days after exposure to a transient increase in $[Ca^{2+}]_i$.

Key words. Ischemia—Hypoxia—Ca^{2+}—Glutamate—Neuronal cell death—Hippocampus—Hypoglycemia—Delayed neuronal cell death

Introduction

Neurons in the central nervous system (CNS) are easily denatured and destroyed by ischemia as a result of cerebral thrombosis or hemorrhage or by deprivation of energy supply including glucose or oxygen (hypoglycemia or anoxia/hypoxia). Because neurons are not newly generated, a certain part of CNS functions are lost irreversibly, granted that they are compensated by surviving neurons with their plastic nature. Although such large-scale CNS damage would not frequently occur, small-scale neu-

[1]Laboratory of Cellular Neurobiology, School of Life Science, Tokyo University of Pharmacy and Life Science, 1432-1 Horinouchi, Hachioji, Tokyo 192-03, Japan
[2]Department of Biology, School of Science, Osaka University, 1-1 Machikaneyama, Toyonaka, Osaka 560, Japan

ronal death could occur rather frequently from a sporadic and localized ischemia. When these small damages accumulate to exceed the ability of plastic compensation, the neurological symptoms appear. Understanding the mechanism(s) of neuronal death after ischemic invasion is, therefore, an urgent demand from both clinical and social points of view.

There is a difference in vulnerability to ischemia among the types of neurons [1,2]. CA1 pyramidal cells in the hippocampus and Purkinje cells in the cerebellum are typically vulnerable neurons, so that amnesia and dyskinesia are representative symptoms in patients of stroke. In contrast, the neurons in the brainstem, the spinal cord, and the peripheral nervous systems are relatively resistant to ischemia. Based on the fact that the neurons with high vulnerability to ischemia are rich in glutamate receptors, an excessive release of glutamate during ischemic or anoxic invasion is hypothesized as a principal cause of neuronal death [3,4].

In this chapter, the glutamate hypothesis of ischemia-induced neuronal death is explained in reference to calcium. We know, however, that the glutamate receptors are distributed most widely in the CNS. Is it reasonable to explain the regional difference in vulnerability between neurons simply by the presence or absence of the glutamate receptors? If so, how does glutamate lead neurons to cell death? We demonstrate here some evidence and discuss the role of glutamate in neuronal cell death.

Acute and Delayed Neuronal Death

When an experimental animal is caused to inhale O_2-free gas or its cerebral circulation is occluded, the level of cellular adenosine triphosphate (ATP) drops quickly. In parallel to that, cell excitability and synaptic transmission are suppressed [5–9]. If this condition continues for a long time, the hippocampal neurons undergo irreversible disintegration, that is, acute neuronal death. If the ischemic invasion is limited to a short period, however, the cellular ATP level is restored and the neuronal functions including membrane potential, membrane resistance, synaptic transmission, and protein synthesis rate recover to almost the same levels as those before the invasion. The vulnerable neurons, however, lose their neuronal function again and die several days later in spite of a transient recovery. This phenomenon is called delayed neuronal death [1,2].

Because the resting membrane potential is built on the concentration gradients of ions across the cell membrane, which are mainly maintained by an ion-pumping function dependent on intracellular ATP, depletion of ATP following ischemia results in depolarization of the membrane potential. The cause of membrane depolarization is further discussed later. Depolarization activates the voltage-gated Na^+ channel, leading to a massive influx of Na^+. This should produce an influx of membrane-permeable anion, Cl^-, from the requirement of electrical neutrality (Donnan equilibrium). Consequently, neurons become hyperosmotic. The difference in osmolality across the cell membrane gives rise to an inflow of water and causes cell swelling, which can physically destroy the cell when hydrostatic pressure exceeds the mechanical strength of the membrane. This process favors the explanation of acute neuronal death by the osmotic hypothesis or edema hypothesis. If the depolarization of neurons and the imbalance of ion distribution are maintained for a period long enough to achieve Donnan equilibrium as just described, neurons undergo acute cell death. It has been observed that extracellular volume was almost halved after ischemic invasion [10].

What, then, is the mechanism of delayed neuronal death? Its mechanism seems different from that of acute cell death because the destruction of the cell membrane is not caused by a brief ishemia, which can induce delayed neuronal death.

Ca^{2+} Overload Hypothesis of Delayed Neuronal Death

Because the cellular functions of neurons apparently recover after a transient and brief ischemia, the osmotic hypothesis does not explain delayed neuronal death. Instead, a hypothesis based on an overload of Ca^{2+} inside the cell is proposed to explain this type of cell death [11]. This hypothesis is that the accumulation of Ca^{2+} inside the cell over the range of physiological concentration activates some Ca^{2+}-dependent mechanism(s) that is (are) not usually activated by physiological $[Ca^{2+}]_i$, leading to cell death.

Several pathways in neurons lead to Ca^{2+} overload during ischemia. The first is the voltage-gated Ca^{2+} channel. There is a great concentration gradient of Ca^{2+} across the cell membrane. The free Ca^{2+} concentration inside the cell ($[Ca^{2+}]_i$) is less than $10^{-7} M$, while that outside the cell ($[Ca^{2+}]_o$) is about $10^{-3} M$. As the concentration gradient is of the order of 10^4, $[Ca^{2+}]_i$ markedly rises to $10^{-6} M$ or higher once the voltage-gated Ca^{2+} channels are activated.

The second pathway to Ca^{2+} overload is the Ca^{2+} extrusion system. This system consists of Na^+/Ca^{2+} exchanger in the plasma membrane, Ca^{2+}/H^+ antiporter in the mitochondrial membrane, and Ca^{2+}/Mg^{2+}-ATPase in the endoplasmic reticular membrane. They all play a role in maintaining low $[Ca^{2+}]_i$. Because they are fueled directly or indirectly by ATP, collapse of energy supply leads to the elevation of $[Ca^{2+}]_i$. Although Na^+/K^+-ATPase is not directly involved in the extrusion of Ca^{2+} from the cells, its decrease results in an accumulation of K^+ ion in the extracellular space, which further depolarizes the membrane as if it were a positive feedback of depolarization. It should be noted that the Na^+/Ca^{2+} exchanger is electrogenic; efflux of one Ca^{2+} ion is coupled to influx of three Na^+ ions [12]. Its reversal potential lies 10–20 mV above the resting membrane potential. Thus, membrane depolarization retards the export of Ca^{2+} and even reverses the direction of antiport to lead to an uptake of Ca^{2+}. Na^+ concentration inside and outside the cell also affect the ability and the direction of ion transport [13]. If the intracellular Na^+ concentration is increased, e.g., by Na^+ influx through voltage- or ligand-gated Na^+ channels, the Na^+/Ca^{2+} exchanger functions so that $[Ca^{2+}]_i$ increases so as to extrude intracellular Na^+.

These conditions occur in presynaptic terminals as well as in postsynaptic neurons. The elevated $[Ca^{2+}]_i$ in the synaptic terminals triggers the release of glutamate, an excitatory transmitter. Released glutamate activates the receptor-coupled ion channels in the postsynaptic membranes, which is the third pathway to elevate $[Ca^{2+}]_i$ and is another cause of membrane depolarization. The foregoing consequence of events explains the observation that the membrane depolarizes self-regeneratively once the energy level falls below a certain critical point. We have also ascertained that $[Ca^{2+}]_i$ increases in a self-regenerative manner [14]. As summarized in Fig. 1, deprivation of oxygen and glucose, exposure to high-K^+ medium, and application of a high dose of

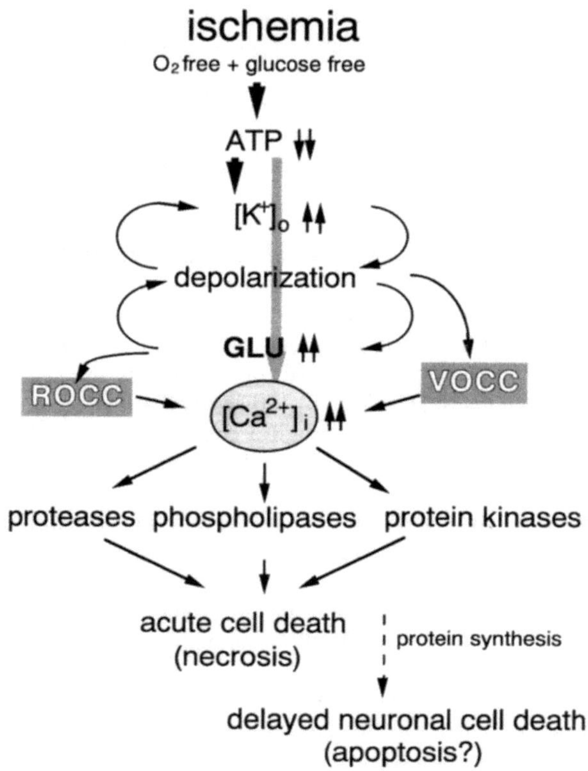

FIG. 1. Mechanism underlying Ca^{2+}-dependent cell death caused by ischemia. *GLU*, glutamate; *ROCC*, receptor-operated Ca^{2+} channel; *VOCC*, voltage-operated Ca^{2+} channel

glutamate may have the equivalent effects and induce $[Ca^{2+}]_i$ increase in fura-2-loaded hippocampal slices.

A number of enzymatic systems and functional proteins are activated by Ca^{2+} [15]. Cells employ Ca^{2+} as an intracellular signal molecule to regulate their physiological functions. However, an excessively elevated $[Ca^{2+}]_i$ overactivates these enzymatic systems to initiate pathogenic processes, including the digestion of cellular components and the breakdown of cell membrane. Protein phosphorylation enzymes such as a Ca^{2+}/diacylglycerol-dependent protein kinase can potentiate the activity of the voltage-gated channels to further enhance the depolarization and $[Ca^{2+}]_i$ elevation [16]. It has been shown that some enzymes upregulate themselves by self-phosphorylation [17]. Thus, some of the primed enzymes remain activated and continue to destroy the cell structure and the functional components if the high $[Ca^{2+}]_i$ state ends after a short period but one long enough to activate Ca^{2+}-dependent enzymes.

Expression of certain species of genes is known to be activated by Ca^{2+}. Among these, some transcription factors have been identified. Although little is known about the proper target of these factors, it is assumed that some key proteins for the long-term enhancement of excitatory synaptic transmission are included in the target [18]. These gene products may act in the direction of cell death rather than in the direction of survival. In this context, attention has been attracted to whether the genes related to apoptosis, a programmed cell suicide process occurring as a part of normal development, are activated by ischemic invasion. It has recently been demonstrated that delayed neuronal death in CA1 pyramidal cells is apoptosis, which is characterized by cell shrinkage and DNA fragmentation [19].

The Ca^{2+} overload hypothesis for delayed neuronal death is supported by several lines of experimental evidence that are shown in the following sections.

Cultured Neurons and Brain Slices as Models for Analysis of Neuronal Death

Appropriate experimental models are required to analyze the mechanism(s) underlying neuronal death. Vessel occlusion models in the gerbil and rat are often employed as an in vivo experimental ischemic model. In the brain in vivo, uneven distribution of blood vessels should be taken into account for region-specific brain damage. An additional cause of heterogeneous vulnerability is the distribution of neuroglia, which relay oxygen and nutrients from blood to neurons. To escape from these structural uncertainties and to analyze the vulnerability of neurons per se, cultured neurons isolated from fetal brain and dispersed in Petri dishes have been utilized. Under proper conditions, the neurons can be maintained in this artificial environment for more than several weeks with or without the coexistence of nonneuronal cells. The cultured neurons, e.g., hippocampal pyramidal cells, are sensitive to exogenously applied glutamate and easily undergo disintegration, serving as a good model system for the analysis of glutamate-induced death [20–22].

The cultured neurons, however, have intrinsic problems. The neurons should be taken from embryonic brains so that their properties, including receptor distribution and enzyme expression, might be nonidentical to the neurons in situ. The cells have inevitably experienced anoxia during the procedures of cell isolation and hypoxia during culture. Relative resistance of cultured neurons to anoxia/hypoxia might be related to this fact. Extremely vulnerable cells would not survive the low-O_2 condition during the isolation process and culture condition. In our preliminary examination, the hippocampal pyramidal cells ramaining in culture could survive a 30-min anoxia. In spite of these problems, the nakedness of cultured cells and the ease of controlling ambient conditions are valuable. In fact, it is in the cultured hippocampal neurons that the glutamate-induced elevation of $[Ca^{2+}]_i$ in neuronal cytoplasm was first demonstrated [23] and that a quantitative comparison between the magnitudes of cell death and $[Ca^{2+}]_i$ increase was first made [22].

Fresh brain slices prepared from adult animals under low temperature can be maintained for as long as 12 h if they are kept in an appropriate medium gassed with oxygen. This preparation allows us to study the changes in synaptic function and cell excitability during and after ischemic invasion. The regional changes in $[Ca^{2+}]_i$ can be also monitored by using Ca^{2+} indicators in parallel with electrical activity [9]. Although a lifetime of 12 h for the slice preparation is not sufficient to pursue the whole process of ischemic damage, it should be possible to detect the early events and the progressive changes that are triggered by ischemia and would cause delayed neuronal death. The results obtained in these in vitro models should eventually be reexamined in the brain in vivo, but they doubtlessly provide us with the clues to elucidate the mechanism(s) of neuronal death.

From ATP Depletion to Membrane Depolarization

It has been demonstrated that the level of ATP in the brain does decrease rapidly after ischemic invasion [5,24]. We mentioned earlier that depletion of ATP causes membrane depolarization during ischemia, called anoxic depolarization. Although the causal relationship between these two phenomena has not been fully understood, activation of the ATP-dependent K^+ channel and retardation of the Na^+/K^+ pump because of the loss of intracellular ATP have been thought to be responsible for the anoxic depolarization. The K^+ ATP channel is activated under the lower ATP concentration inside the cells, while it is closed in the presence of ATP [25]. Intracellular recordings in hippocampal CA1 pyramidal cells have revealed slight hyperpolarization immediately after the ischemic condition, followed by a large depolarization several minutes afterward [6,8]. During ischemia, the K^+ ATP channel is activated by the decrease of intracellular ATP, resulting in efflux of K^+ ion and hyperpolarization to the equilibrium potential of K^+ ion. Consequently, K^+ ion accumulates in the extracellular space because of the dysfunction of the Na^+/K^+ pump to uptake K^+ ion into neurons caused by the loss of ATP. Elevated $[K^+]_o$ depolarizes around neurons according to Nernst's equation, and voltage-dependent channels including voltage-dependent Ca^{2+} channels are activated, which then leads to the large depolarization of neurons following the early hyperpolarization. Involvement of the K^+ ATP channel and the Na^+/K^+ pump in the change of membrane potential is supported by the facts that hyperpolarization is blocked by sulfonylureas, K^+ ATP channel blockers, and that depolarization can be induced by ouabain, an inhibitor of the Na^+/K^+ pump. It should be taken into account that energy failure occurs in glial cells as well. Because the glial cells take up extracellular glutamate and K^+ ion released from neurons in an energy-dependent manner, glial ischemia should hold the levels of glutamate and K^+ high in the extracellular space; both of these enhance the depolarization of neurons.

From Depolarization to Glutamate Release: Is Glutamate Release a Result or a Cause of Neuronal Death?

Extracellular concentration of glutamate does increase during ischemia, as monitored by brain microdialysis [26,27]. Glutamate, however, is a common amino acid that constitutes part of some proteins, so that the increased extracellular glutamate might be not a cause but a result of neuronal death. We thus examined the effects of exogenously applied glutamate in cultured neurons to investigate if it actually leads to neuronal death [22]. When glutamate was administered for 15 min (Fig. 2), the hippocampal cells degenerated and the number of neurons markedly decreased in 24 h. The toxic effect of glutamate was dependent on its dose and the duration of application (Fig. 3a,b). Toxicity was greatly enhanced when $[Ca^{2+}]_o$ was raised to more than 3 m*M* (Fig. 3c). Addition of 3 m*M* Mg^{2+} reduced glutamate toxicity (Fig. 3d). Toxicity was reproduced by agonists specific for receptor subtypes. Quisqualate and kainate exhibited stronger toxicity than glutamate and *N*-methyl-D-aspartate (NMDA), when compared at the same concentration (Fig. 3f). As discussed later, the extent of toxicity corresponded well to that of the $[Ca^{2+}]_i$ elevation (Fig. 4).

Cell death of cultured neurons seemed to progress gradually, and no obvious segregation between the acute and the delayed cell deaths was recognized over the

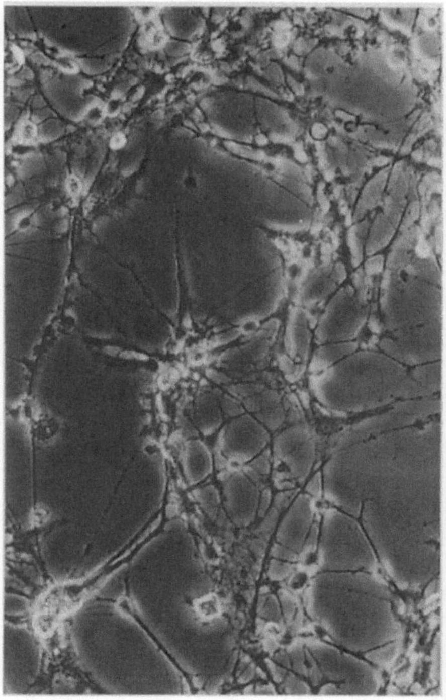
a

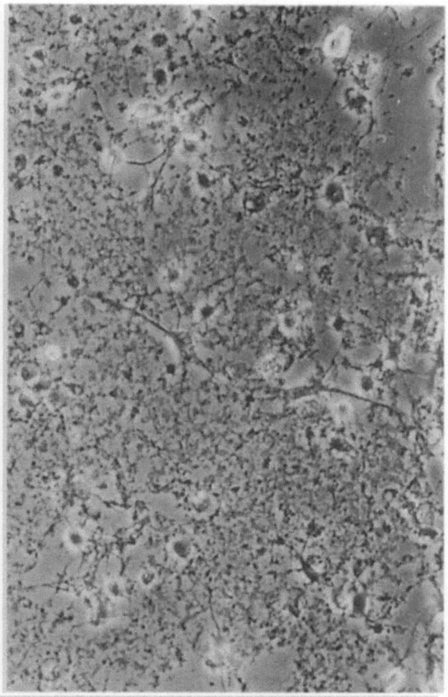
b

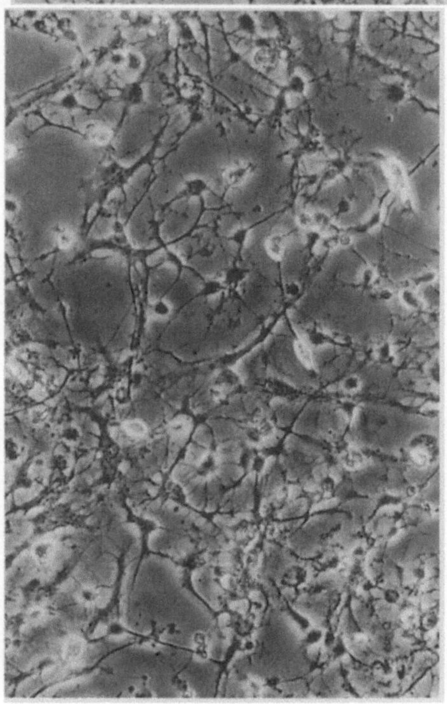
c

FIG. 2a–c. Photomicrographs of cultured hippocampal neurons 24 h after exposure to glutamate. Neurons dissociated from 18-day rat embryo hippocampus were cultured for 7 days and exposed to glutamate ($10^{-3} M$) for 15 min. **a** Control: neuronal cells exposed to medium without glutamate. **b** Cells exposed to medium containing $10^{-3} M$ glutamate. **c** Cells treated with rhod-2/acetoxymethyl-ester (AM) ($5 \mu M$) 1 h before glutamate administration

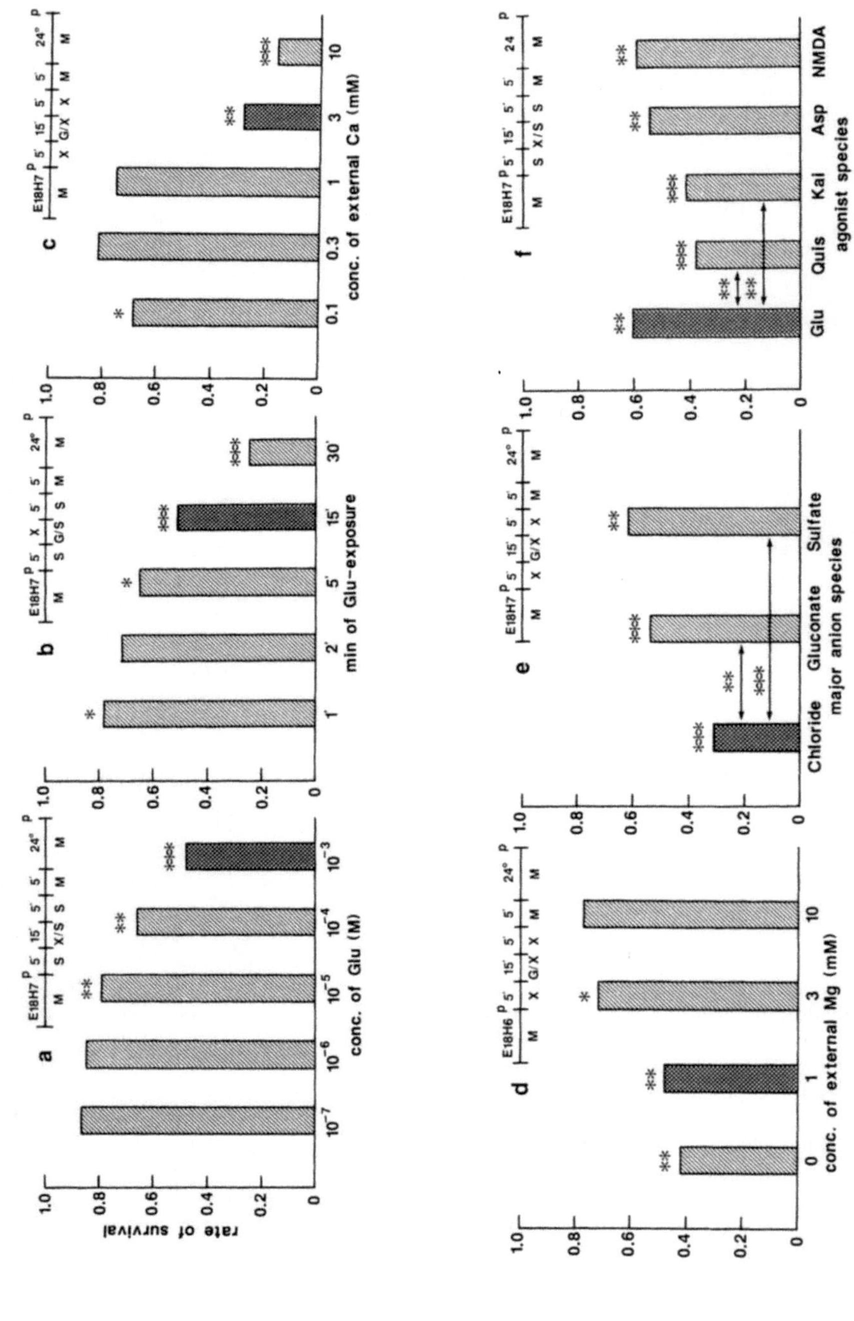
a
b
c
d
e
f
rate of survival
conc. of Glu (M)
min of Glu-exposure
conc. of external Ca (mM)
conc. of external Mg (mM)
Chloride Gluconate Sulfate
major anion species
Glu Quis Kai Asp NMDA
agonist species

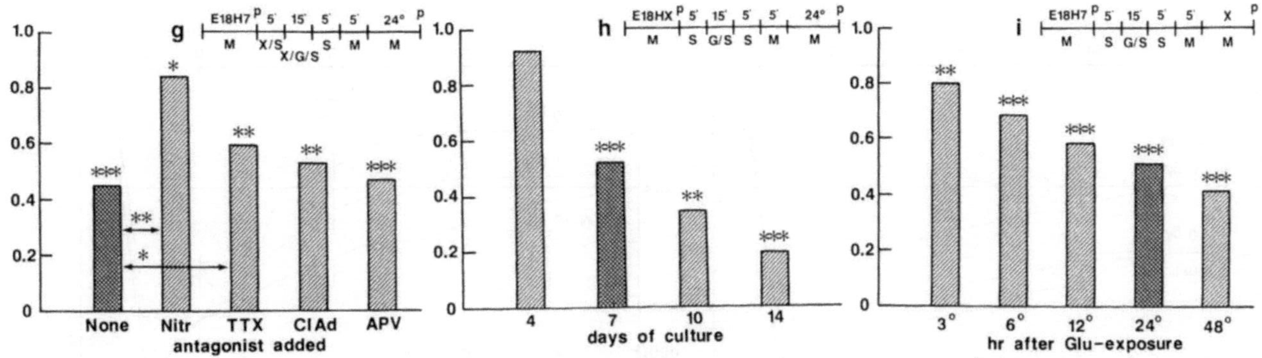

FIG. 3a–i. Quantified neuronal loss after exposure to glutamate (*Glu*): effects of alterations in concentration of glutamate (**a**), duration of exposure to glutamate (**b**), concentration of external $CaCl_2$ (**c**), concentration of external $MgCl_2$ (**d**), species of major anion in the medium (**e**), species of excitatory amino acids (**f**), species of drug delivered concurrently with glutamate (**g**), numbers of days in culture (**h**), interval between exposures to glutamate, and scoring of live cells (**i**). *Quis*, quisqualate; *Kai*, kainate; *Asp*, aspartic acid; *NMDA*, *N*-methyl-D-aspartate. *Doubly hatched column* in each panel signifies standardized condition. Each panel includes a diagram of the experimental procedures in each series examined in which *M*, *S*, *G*, *p*, and *X* indicate culture medium, basal salt solution, glutamate, photographing, and variable, respectively. Error bars are not indicated because each *rate of survival (RS)* in the histogram represents a value of the mean rate of remainder (*RR*) of glutamate-treated specimens divided by mean RR of glutamate-untreated specimens. For some examinations, the Cochran-Cox modified *t*-test for two sample groups of nonidentical variances showed a statistically significant difference, indicated at the top of each column by *($P < .1$), **($P < .05$), or ***($P < .01$). (From [22], with permission)

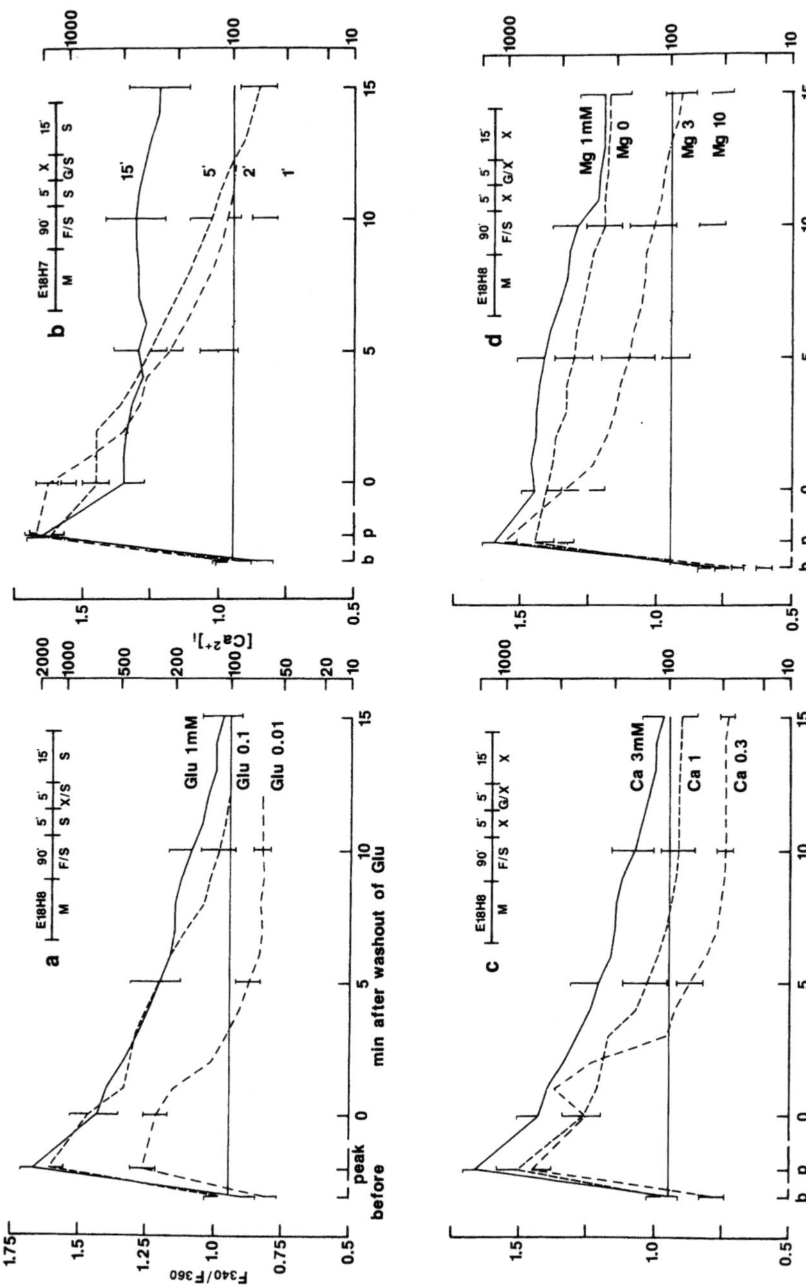
a
b
c
d
E18H8
E18H7
M
F/S
90'
5'
15'
Glu 1mM
Glu 0.1
Glu 0.01
Ca 3mM
Ca 1
Ca 0.3
Mg 1 mM
Mg 0
Mg 3
Mg 10
F340/F360
[Ca2+]i
before
peak
min after washout of Glu

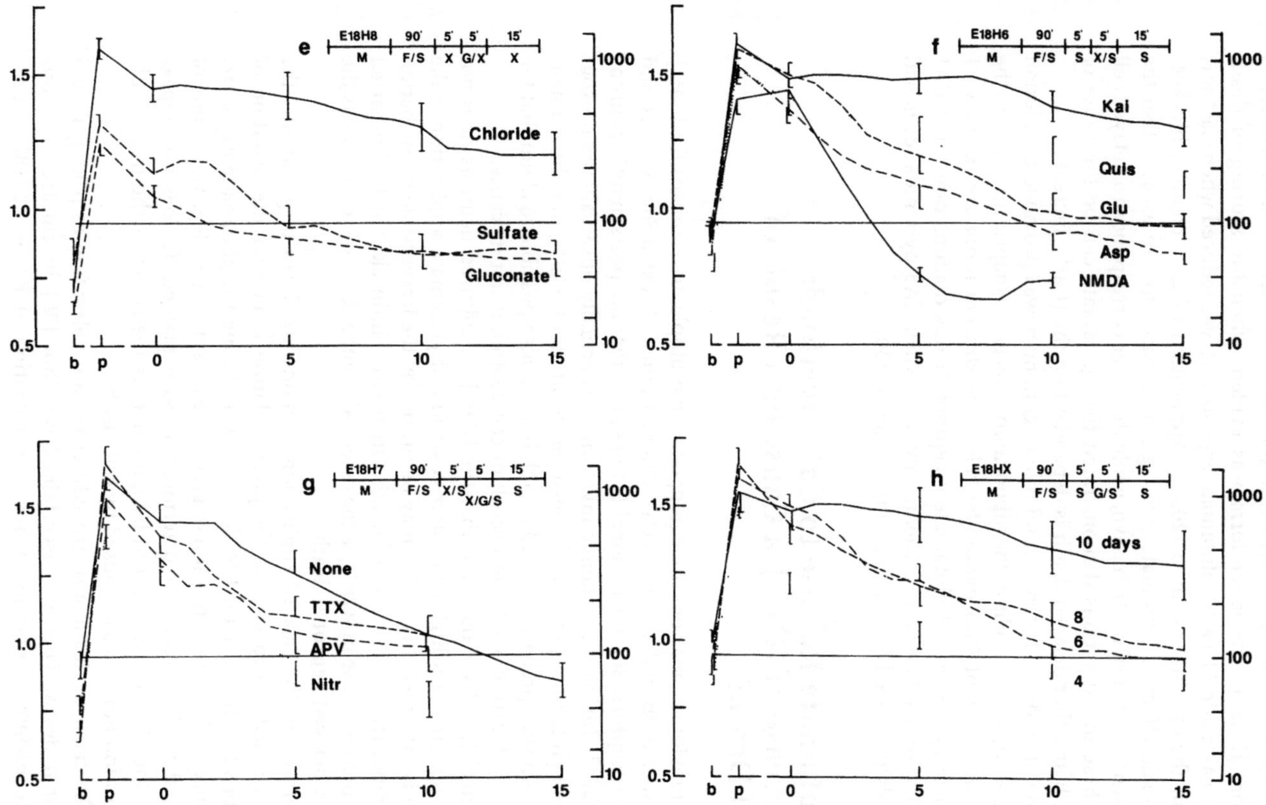

Fig. 4a–h. Quantified cytoplasmic Ca^{2+} concentration ($[Ca^{2+}]_i$) on exposure to glutamate and after its withdrawal: Effects of alterations in concentration of glutamate (**a**), duration of exposure to glutamate (**b**), concentration of external $CaCl_2$ (**c**), concentration of external $MgCl_2$ (**d**), species of major anion in the medium (**e**), species of excitatory amino acids (**f**), species of drugs delivered concurrently with glutamate (**g**), and number of days under culture (**h**). *Left ordinate* shows the ratios of the fluorescence intensities, which are directly correlated to levels of $[Ca^{2+}]_i$ (*right ordinate*). *Bars* indicating standard errors of the means are given at limited timepoints for graphic clarity. *Horizontal* line indicates the 100 n*M* level in $[Ca^{2+}]_i$. Each panel contains a diagrammed experimental paradigm as in Fig. 3 (From [22], with permission)

time-course (Fig. 3i). However, there still appeared to be two forms of cell death mechanism, one Ca^{2+} independent and one Ca^{2+} dependent. The former form was evident 30 min after the glutamate exposure and was prevented by a withdrawal of Cl^- from the external medium. This early death was not reproduced by an exposure to Ca^{2+} ionophore [21,28]. The latter form was evident when the scoring of dead/survived cells was made 24 h after glutamate exposure. It was reduced when Ca^{2+} was removed from the external medium [20,22]. Application of Ca^{2+} channel blockers reduced the toxic effect of glutamate (Fig. 3g). It is reasonable to assume that the former represents acute cell death in vivo, while the latter corresponds to delayed cell death [29]. These investigations demonstrated that glutamate can be the cause of neuronal death and that delayed death is dependent on the $[Ca^{2+}]_i$ increase.

Another line of evidence for the causal role of glutamate was provided by a lesion experiment. CA1 pyramidal cells in the hippocampus receive synaptic inputs from the terminals of the axons of CA3 pyramidal cells. The *stratum radiatum* in the CA1 region is rich in synapses and glutamate receptors. It was demonstrated that CA1 pyramidal cells survived a severe ischemic invasion when CA3 pyramidal cells had been previously destroyed by pharmacological means [30].

From Glutamate Release to Ca^{2+} Overload: Is the Increase of $[Ca^{2+}]_i$ a Cause or a Result of Neuronal Death?

As described previously, neurons are equipped with multiple mechanisms for $[Ca^{2+}]_i$ mobilization such as the voltage-gated Ca^{2+} channels, the Ca^{2+} extrusion systems, and the glutamate receptors. So, it is reasonable to assume that the postsynaptic neurons increase in $[Ca^{2+}]_i$ during the ischemic invasion. It is actually observable that exogenously applied glutamate induces a large-scale elevation of $[Ca^{2+}]_i$ in cultured neurons and in brain slice preparations [23,31,32], but it is also possible to claim that the observed $[Ca^{2+}]_i$ elevation would be a result of cell death but not its cause.

We and others [33] have observed that the $[Ca^{2+}]_i$ in cultured neurons does not decrease and remains at a high level after cessation of glutamate application if the concentration is extremely high. This may be caused by the irreversible perturbation of the cell membrane, that is, acute cell death. Otherwise, Limbrick et al. [33] reported that this sustained increase in $[Ca^{2+}]_i$ is the essential event of Ca^{2+} overload and is the major cause of delayed neuronal death.

To the contrary, in the case of application of moderate doses of glutamate, the $[Ca^{2+}]_i$ level eventually recovered to the prestimulation level after the cessation of glutamate exposure, and it tended to decrease even during the glutamate exposure, possibly because of the desensitization of glutamate receptors (see Fig. 4). In spite of this recovery, the cells underwent deterioration, as explained. It does not appear plausible that the $[Ca^{2+}]_i$ rise induced by glutamate is the result of the irreversible perturbation of ion homeostasis caused by cell death.

If Ca^{2+} is the cause of death, any procedures to cause elevation of $[Ca^{2+}]_i$ should induce neuronal death. This was ascertained by Choi [28] in cultured neurons by applying a medium containing a high concentration of K^+ or A23187, a Ca^{2+} ionophore.

Our experiments also support this result. We loaded neurons with various types of Ca^{2+} chelators to suppress the $[Ca^{2+}]_i$ elevation. As shown in Fig. 2c, the cultured hippocampal cells preloaded with rhod-2, which is a Ca^{2+} indicator with the Kd value of 1000 n*M* for Ca^{2+}, showed a significant resistance to glutamate. It is interesting to add a comment here that the Ca^{2+} chelators effective to resist the glutamate toxicity appeared to have an optimal range of binding constant with Ca^{2+}. O,O′-Bis(2-aminophenyl)ethyleneglycol-N,N,N′,N′-tetraacetic acid (BAPTA), with a very high affinity to Ca^{2+}, showed its own neurotoxic effect when survival rate was scored 24 h after glutamate exposure. Quin-2, another Ca^{2+} indicator, also showed a protective effect in CNS preparations in vivo. When frog spinal motoneurons were previously loaded with quin-2, the cells acquired tolerance to the application of a high dose of glutamate or of A23187 [34].

In summary, the facts that the increase in $[Ca^{2+}]_i$ induces cell death and that manipulation to suppress the increase in $[Ca^{2+}]_i$ inhibits cell death indicate that the increase in $[Ca^{2+}]_i$ is involved in the neuronal cell death induced by glutamate application.

From Glutamate Release to Ca^{2+} Overload: Endogenous Ca^{2+} Buffers Inside Neurons

In this context it is interesting to consider the role of the endogenous Ca^{2+} chelators inside neurons. There is a species of CNS-specific Ca^{2+}-binding protein, called calbindin D-28K (CB28), whose function is unspecified in spite of its abundant presence in some types of neurons.

Freund et al. [35] reported that neurons with high immunoreactivity to an anti-CB28 antibody survived through the ischemic invasion. They interpreted this result to be that the neurons containing a high level of CB28 were selectively resistant to the glutamate-induced Ca^{2+} overload. There is another interpretation, however; the Ca^{2+} loading per se might induce the expression of CB28, so that the surviving neurons showed high CB28 immunoreactivity.

We examined the relationship between the behavior of $[Ca^{2+}]_i$ increase and CB28 immunoreactivity in individual neurons. Some of the results are shown in Fig. 5; CB28-positive neurons showed faster recovery from the high $[Ca^{2+}]_i$ state than did CB28-negative neurons. The equivalent result was reported by others [36]. It has not been determined whether CB28 functions as a simple Ca^{2+} chelator. The faster recovery is apparently contrary to the theory of buffer action; a buffer should slow the rate of both the increase and decrease of $[Ca^{2+}]_i$. Further investigation is necessary to know whether CB28 has a yet-unidentified function to accelerate the extrusion of Ca^{2+}.

It is reported that the brains of Alzheimer's disease patients show a reduced CB28-immunoreactivity [37]. It may be that the Ca^{2+}-binding proteins of this class, CB28 or its related proteins such as calretinin, play some determinative roles in brain damage and neuronal death.

From Glutamate Release to Ca^{2+} Overload: Is Any Subtype of Glutamate Receptor Responsible for $[Ca^{2+}]_i$ Elevation?

Glutamate receptors are classified into ionotropic and metabotropic types. Among ionotropic receptors are NMDA and non-NMDA subtypes. From electrophysiological examinations, the NMDA receptor has been nominated as the sole subtype of

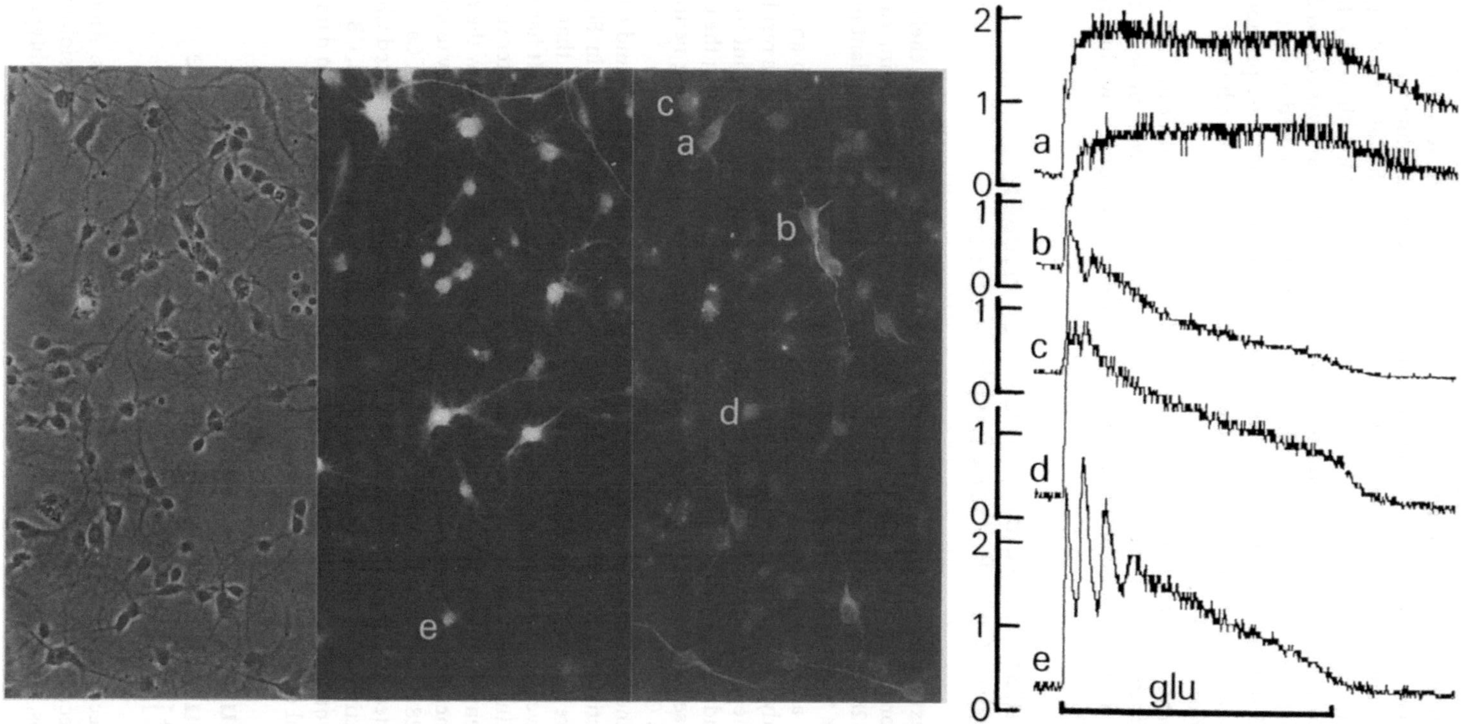

Fig. 5. $[Ca^{2+}]_i$ changes induced by glutamate (*glu*) in cultured hippocampal neurons: photomicrographs with phase-contrast optics (*left*), with anti-CB28 antibody staining (*middle*), and with anti-MAP2 antibody staining (*right*). Recordings in the *right-hand panel* show the $[Ca^{2+}]_i$ change in cells indicted in micrographs: *a* and *b*, CB28-negative neurons; *c* and *d*, CB28-positive neurons; *e*, a CB28-positive nonneuronal cell. Glutamate (10^{-4} *M*) was administered during the *barred* period (160 s). *Ordinates* indicate $[Ca^{2+}]_i$ ($\times 10^{-7}$ *M*)

glutamate receptor that permeates the Ca^{2+} ion [38]. Many recent researches on glutamate toxicity take this assumption a priori, and much effort has concentrated on finding novel NMDA receptor antagonists as antidementia remedies. However, non-NMDA receptor agonists manifests toxic effects as well as NMDA receptor agonists (for cultured cells, see Fig. 3f; for brain slice, see [39]). Non-NMDA receptor antagonists are as effective as NMDA-receptor antagonists in suppressing the glutamate-induced death of cultured neurons [40].

We have demonstrated that some non-NMDA receptors are permeable to Ca^{2+} [41,42]. We adopted Ca^{2+} fluorometry as a more direct methodology than electrophysiology for the examination of Ca^{2+} permeability of the non-NMDA receptor, because ionic currents detected by electrodes do not specify the species of permeating ion. As had been expected, non-NMDA-receptor agonists induced a large elevation in $[Ca^{2+}]_i$ of cultured hippocampal neurons (Fig. 4f). To exclude the contribution of the voltage-gated Ca^{2+} channels in elevating the $[Ca^{2+}]_i$, we repeated the stimulation of non-NMDA receptor in the presence of blockers of the voltage-gated channels. As shown in Fig. 6, a non-NMDA-receptor agonist, kainate, produced a large increase in

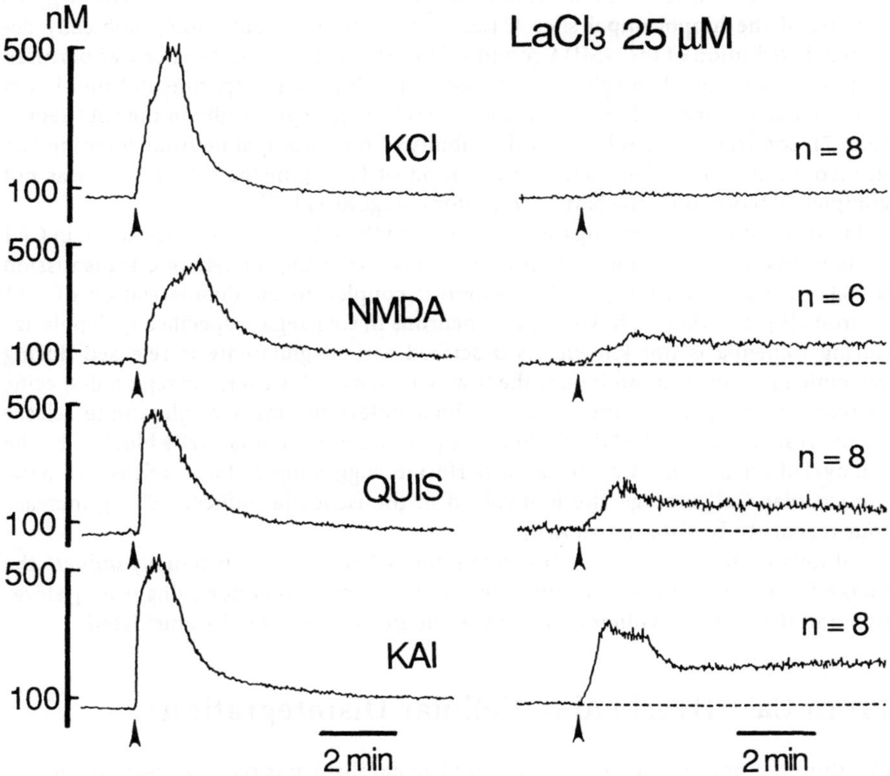

Fig. 6. Effects of glutamate receptor subtype agonists on hippocampal neurons in culture before and after La^{3+} treatment. *KCl*, *N*-methyl-D-aspartate (*NMDA*), quisqualate (*QUIS*), or kainate (*KAI*) was applied to the same populations of neuron (separate dishes were used for different drugs). Each response is the result of digital averaging over 6–8 neurons (From [31], with permission)

$[Ca^{2+}]_i$ in the presence of tetrodotoxin and La^{3+} as an Na^+ and a Ca^{2+} channel blocker, respectively.

The $[Ca^{2+}]_i$ rise mediated by the non-NMDA receptor was dependent on the presence of external Ca^{2+} and was not suppressed by NMDA-receptor antagonists, indicating that the non-NMDA receptor could couple with an ionic channel permeable to Ca^{2+} [41]. In detailed electrophysiological reexamination, Iino et al. [42] reached the identical conclusion and designated this Ca^{2+}-permeable non-NMDA receptor as a type 2 non-NMDA receptor. Recent molecular biological analysis has demonstrated that several clones encoding non-NMDA receptors are permeable to Ca^{2+} and that the combination of isoforms forming an oligomeric assembly determines the Ca^{2+} permeability [43].

Metabotropic glutamate receptor [44] also induces the increase in $[Ca^{2+}]_i$ by accelerating phosphoinositide turnover. The activation of any subtypes of glutamate receptor should result in the elevation of $[Ca^{2+}]_i$ irrespectively of its subtypes. When a hippocampal slice was exposed to NMDA- and non-NMDA-receptor agonists, the $[Ca^{2+}]_i$ rise was induced by any of the agonists in a region-specific manner, presumably representing a differential distribution of receptor subtypes [31,32].

It should be pointed out that NMDA elevated the $[Ca^{2+}]_i$ in CA1 and dentate gyrus regions of the hippocampal slice [31,32], confirming the autoradiographically depicted distribution of the NMDA receptor [45]. However, when the slice was perfused with oxygen-depleted and glucose-free medium, which is an experimental simulation of ischemia, a marked elevation of $[Ca^{2+}]_i$ was induced specifically in the CA1 region (Fig. 7), confirming the selective vulnerability of hippocampal neurons identified by in vivo studies [1,2]. This selective elevation of $[Ca^{2+}]_i$ in the CA1 region was not completely blocked by the NMDA-receptor antagonists [27].

Many questions concerning the mechanism of the selective $[Ca^{2+}]_i$ elevation in CA1 remain unanswered. Simultaneous extracellular recording of synaptic transmission in CA1 revealed that the $[Ca^{2+}]_i$ elevation is coupled to the depolarization of CA1 neurons [9]. The reason, however, why neurons in CA1 region specifically depolarize during ischemia is not known. As described earlier, glutamate is released during ischemic invasion and can induce the $[Ca^{2+}]_i$ increase. However, the regional-specific increase in the $[Ca^{2+}]_i$ cannot be explained unless the massive glutamate release occurs restrictively in the CA1 region. The $[Ca^{2+}]_i$ elevation was partly blocked by the removal of extracellular Ca^{2+} from the perfusate, suggesting that Ca^{2+} release from the intracellular Ca^{2+} storage site is involved in the ischemia-induced $[Ca^{2+}]_i$ increase (our recent study; data not shown).

Although it has been demonstrated that the ischemic invasion actually induces the marked $[Ca^{2+}]_i$ increase in the hippocampus, the mechanism underlying $[Ca^{2+}]_i$ elevation and the selective vulnerability of CA1 neurons remains to be elucidated.

From Ca^{2+} Overload to Cellular Disintegration

In cultured hippocampal neurons, a complete parallelism is seen between the degree of cell death and the magnitude of $[Ca^{2+}]_i$ elevation (compare Fig. 3 with Fig. 4; Ogura et al. [22]). Which is more critical to initiate the death process, the height of peak $[Ca^{2+}]_i$ or the duration of the high $[Ca^{2+}]_i$ state? Because of inactivation of the Ca^{2+} entry pathways or activation of the Ca^{2+} extrusion systems, $[Ca^{2+}]_i$ tended to decrease after peaking at about 2 min, even in the presence of receptor agonist. Peak $[Ca^{2+}]_i$ in

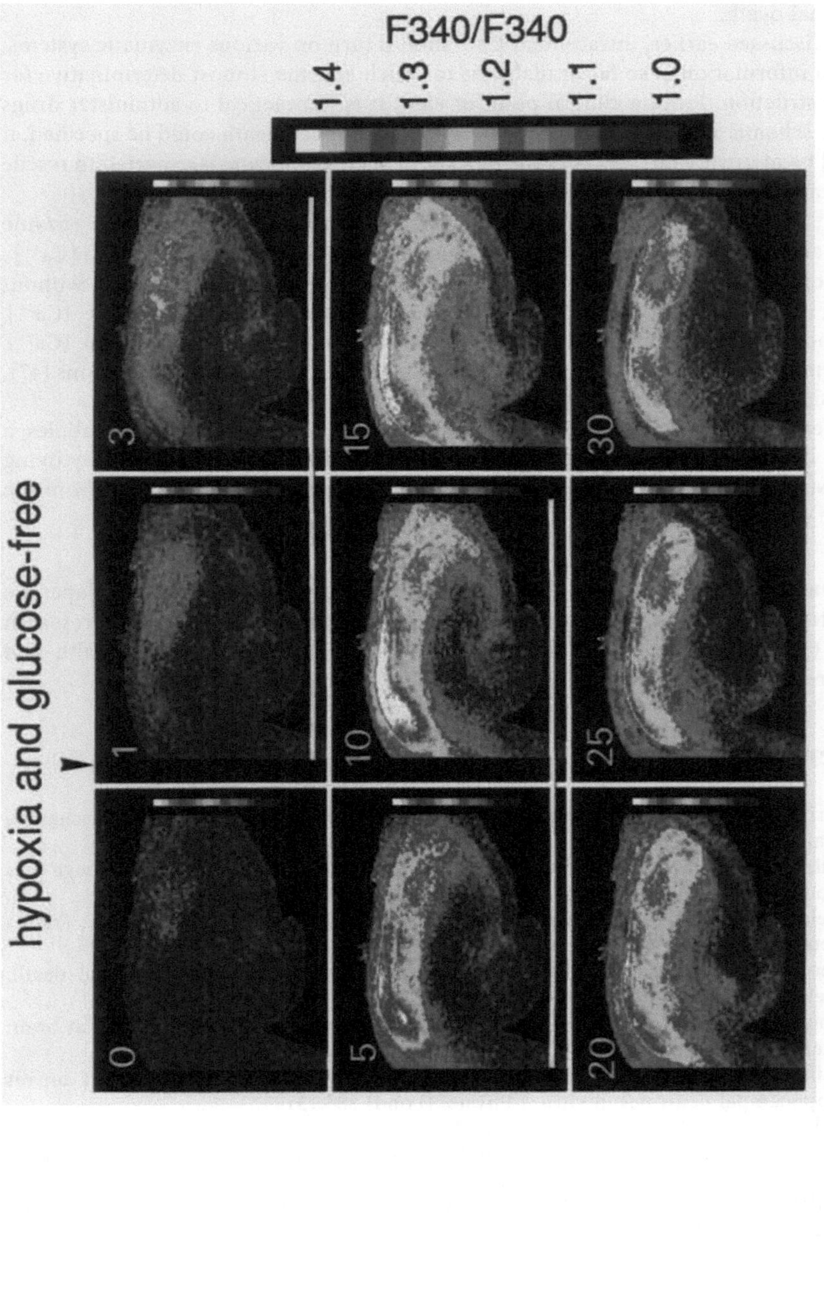

FIG. 7. Changes in $[Ca^{2+}]_i$ during the exposure to an oxygen-free/glucose-free medium for 10 min. *Number* in each panel indicates time after the exposure in minutes; *White line* indicates the continuation of anoxia. Experimental methods have been described in our previous papers. [31,32]

the case of 5 or 15 min of exposure was, perhaps, similar to that in the case of a 2-min exposure, although the absolute level of peak $[Ca^{2+}]_i$ was difficult to determine because fluorometry with fura-2 is only valid in the range of about 10–1000 nM because of the indicator's binding constant with Ca^{2+}. Nevertheless, the degree of cell death was greater in the case of longer exposure. This result suggests that the duration of the high $[Ca^{2+}]_i$ state is more critical to the activation of the cellular process inducing neuronal death.

As discussed earlier, intracellular Ca^{2+} should turn on various enzymatic systems, but no information is so far available as to which enzyme is most determinative for cell destruction. From a clinical point of view, it is impractical to administer drugs before schemic invasion. If any enzyme(s) critical for cell death could be specified, it would be of great value, because the blocker of such an enzyme is expected to rescue the death-destined cells by applying it even after the invasion.

Ca^{2+}-activated neutral protease is one of the candidate enzymes. This enzyme originally has a low affinity to Ca^{2+}, but once it is activated by a surge of high $[Ca^{2+}]_i$, it decomposes. This autolysis greatly alters the enzyme's affinity to Ca^{2+} without losing its proteolytic activity. Therefore Ca^{2+} acts as a switch for the enzyme; $[Ca^{2+}]_i$ does not need to be continuously high [46]. So far the processes following the $[Ca^{2+}]_i$ rise remain unclear. Because the cytocidal effect of Ca^{2+} is not limited to neurons [47], analysis using nonneuronal cells may provide a breakthrough.

Whether or not related to cell death, Ca^{2+} is known to disassemble microtubules, a major cytoskeletal component [48]. Morphological disintegrations shown by dying cells, such as vesicularization (bead formation) of dendrite, should result from the failure of the cytoskeleton.

Acknowledgments. This work was supported by a grant-in-aid from the Japanese Ministry of Education, Science, Sports and Culture (no. 01659002) and the research grant (1A1) for nervous and mental disorders from the Ministry of Health and Welfare.

References

1. Kirino T (1982) Delayed neuronal death in the gerbil hippocampus following ischemia. Brain Res 239:57–69
2. Pulsinelli WA, Brierly JB, Plum F (1982) Temporal profile of neuronal damage in a model of transient forebrain ischemia. Ann Neurol 11:499–509
3. Meldrum B (1985) Excitatory amino acids and anoxic/ischaemic brain damage. Trends Neurosci 8:47–48
4. Choi DW (1995) Calcium: still center-stage in hypoxic-ischemic neuronal death. Trends Neurosci 18:58–60
5. Onodera H, Iijima K, Kogure K (1986) Mononucleotide metabolism in the rat brain after transient ischemia. J Neurochem 46:1704–1710
6. Fujiwara N, Higashi H, Shimoji K, Yoshimura M (1987) Effects of hypoxia on rat hippocampal neurones in vitro. J Physiol (Lond) 384:131–151
7. Sick TJ, Solow EL, Roberts EL Jr (1987) Extracellular potassium ion activity and electrophysiology in the hippocampal slice: paradoxical recovery of synaptic transmission during anoxia. Brain Res 418:227–234
8. Rader RK, Lanthorn TH (1989) Experimental ischemia induces a persistent depolarization blocked by decreased calcium and NMDA anagonists. Neurosci Lett 99:125–130

9. Nakamura T, Akita K, Katayama Y, Sasamoto K, Kudo Y (1990) Preferential elevation of intracellular Ca^{2+} at CA1 region during anoxia in rat hippocampal slice as revealed by rhod-2 fluorometry. Jpn J Pharmacol 52:190P
10. O'Brien MD (1979) Ischemic brain edema. A review. Stroke 10:623–628
11. Siesjo K, Wieloch T (1985) Brain ischemia and cellular calcium homeostasis. In: Godfraind T, Vanhoutte PM, Govoni S, Poletti R (eds) Calcium entry blockers and tissue protection. Raven, New York, pp 139–149
12. Kimura J, Miyamae S, Noma A (1987) Identification of sodium-calcium exchange current in single ventricular cells of guinea pig. J Physiol (Lond) 384:199–222
13. White RJ, Raynolds IJ (1995) Mitochondria and Na^{+}/Ca^{2+} exchange buffer glutamate-induced calcium loads in cultured cortical neurons. J Neurosci 15:1318–1328
14. Kudo Y, Ogura A, Sugita S, Higashi H (1988) Visualization of hypoxia-induced cytosolic calcium elevation in hippocampal neurons. Abstr Soc Neurosci 14:186
15. Campbell AK (1983) Intracellular calcium: its universal role as regulator. Wiley, Chichester
16. Adamson P, Hajimohamadreza I, Brammer MJ, Campbell IC (1989) Intra-synaptosomal free calcium concentration is increased by phorbol esters via a 1,4-dihydropyridine-sensitive (L-type) Ca^{2+} channel. Eur J Pharmacol 162:59–66
17. Miller SG, Kennedy MB (1986) Regulation of brain type II Ca^{2+}/calmodulin-dependent protein kinase by autophosphorylation. A Ca^{2+}-triggered molecular switch. Cell 44:861–870
18. Cole AJ, Saffen DW, Baraban JM, Worley PF (1989) Rapid increase of an immediate early gene messenger RNA in hippocampal neurons by synaptic NMDA receptor activation. Nature 340:474–476
19. Nitatori T, Sato N, Waguri S, Karasawa Y, Araki H, Shibanai K, Kominami E, Uchiyama Y (1995) Delayed neuronal death in the CA1 pyramidal cell layer of the gerbil hippocampus following transient ischemia is apotosis. J Neurosci 15:1001–1011
20. Choi DW (1985) Glutamate neurotoxicity in cortical cell culture is calcium dependent. Neurosci Lett 58:293–297
21. Rothman SM (1985) The neurotoxicity of excitatory amino acids is produced by passive chloride influx. J Neurosci 5:1483–1489
22. Ogura A, Miyamoto M, Kudo Y (1988) Neuronal death in vitro: parallelism between survivability of hippocampal neurones and sustained elevation of cytosolic Ca^{2+} after exposure to glutamate receptor agonist. Exp Brain Res 73:447–458
23. Kudo Y, Ogura A (1986) Glutamate-induced increase in intracellular Ca^{2+} concentration in isolated hippocampal neurones. Br J Pharmacol 89:191–198
24. Lipton P, Whittingham TS (1982) Reduced ATP concentration as a basis for synaptic transmission failure during hypoxia in the in vitro guinea pig hippocampus. J Physiol 325:51–65
25. Noma A (1983) ATP-regulated K^{+} channels in cardiac muscle. Nature 305:147–148
26. Hagberg H, Lehmann A, Sandberg M, Nystrom B, Jacobson I, Hamberger A (1985) Ischemia-induced shift of inhibitory and excitatory amino acids from intra- to extracellular compartments. J Cereb Blood Flow Metab 5:413–419
27. Mitani A, Kadoya F, Nakamura Y, Kataoka K (1990) Visualization of hyoxia-induced glutamate release in gerbil hippocampal slice. Neurosci Lett 122:167–170
28. Choi DW (1987) Glutamate neurotoxicity in cortical cell culture. J Neurosci 7:357–368
29. Koh J-Y, Gwag BJ, Lobner D, Choi DW (1995) Potentiated necrosis of cultured cortical neurons by neurotrophins. Science 268:573–575
30. Onodera H, Sato G, Kogure K (1986) Lesions to Schaffer collaterals prevent ischemic death of CA1 pyramidal cells. Neurosci Lett 68:169–174
31. Kudo Y, Ito E, Ogura A (1991) Heterogeneous distribution of glutamate receptor subtypes in hippocampus as revealed by calcium fluorometry. Adv Exp Med Biol 287:431–440
32. Kudo Y, Nakamura T, Ito E (1991) A "macro" image analysis of fura-2 fluorescence to visualize the distribution of functional glutamate receptor subtypes in hippocampal slices. Neurosci Res 12:412–420

33. Limbrick DD Jr, Churn SB, Sombati S, DeLorenzo RJ (1995) Inability to restore resting intracellular calcium levels as an early indicator of delayed neuronal death. Brain Res 690:145–156
34. Kudo Y, Takeda K, Yamazaki K (1990) Quin2 protects against neuronal cell death due to Ca^{2+} overload. Brain Res 528:48–54
35. Freund TF, Buzsaki G, Leon A, Baimbridge KG, Somogyi P (1990) Exp Brain Res 83:55–66
36. Mattson MP, Rychlik B, Chu C, Christakos S (1991) Evidence for calcium-reducing and excito-protective roles for calcium-binding protein calbindin-D28K in cultured hippocampal neurons. Neuron 6:41–51
37. Iacopino AM, Christakos S (1990) Specific reduction of calcium-binding protein (25-kilodalton calbindin-D) gene expression in aging and neurodegenerative diseases. Proc Natl Acad Sci USA 87:4078–4082
38. McDermott AB, Mayer ML, Westbrock GL, Smith SJ, Barker JL (1986) NMDA-receptor activation increases cytoplasmic calcium concentration in cultured spinal cord neurones. Nature 321:519–522
39. Garthwaite G, Garthwaite J (1991) AMPA neurotoxicity in rat cerebellar and hippocampal slices: histological evidence for three mechanisms. Eur J Neurosci 3:715–728
40. Frandsen A, Drejen J, Schousboe A (1989) Direct evidence that excitotoxicity in cultured neurons is mediated via *N*-methyl-D-aspartate (NMDA) as well as non-NMDA receptors. J Neurochem 53:297–299
41. Ogura A, Akita K, Kudo Y (1990) Non-NMDA receptor mediates cytoplasmic Ca^{2+} elevation in cultured hippocampal neurones. Neurosci Res 9:103–113
42. Iino M, Ozawa S, Tsuzuki K (1990) Permeation of calcium through excitatory amino acid receptor channels in cultured rat hippocampal neurones. J Physiol (Lond) 424:151–165
43. Hume RI, Dingledine R, Heinemann SF (1991) Identification of a site in glutamate receptor subunits that controls calcium permeability. Science 253:1028–1031
44. Sugiyama H, Ito I, Hirono C (1987) A new type of glutamate receptor linked to inositol phospholipid metabolism. Nature 325:531–533
45. Cotman CW, Monaghan DT, Ottersen OP, Storm-Mathisen J (1987) Anatomical organization of excitatory amino acid receptors and their pathways. Trends Neurosci 10:273–280
46. Suzuki K, Tsuji S, Ishida S, Kimura Y, Kubota S, Imahori K (1981) Autolysis of calcium-activated neutral protease of chicken skeletal muscle. J Biochem (Tokyo) 90:1787–1793
47. Schanne FAX, Kane AB, Young EF, Farber JL (1979) Calcium dependence of toxic cell death: a final common pathway. Science 206:700–702
48. Nishida E, Sakai H (1977) Calcium sensitivity of microtubules reassembly system. J Biochem (Tokyo) 82:303–306

Effects of Glutamate Agonists and Antagonists on Ischemic Damage of Hippocampal Neurons

Eiichiro Tanaka, Satoshi Yamamoto, and Hideho Higashi

Summary. Deprivation of both oxygen and glucose produces a transient hyperpolarization and a subsequent slow depolarization in hippocampal CA1 pyramidal neurons in slice preparations. The slow depolarization is followed by a rapid depolarization after approximately 6 min of deprivation. Depolarization reaches −15 mV, and then the membrane potential slowly rises to 0 mV (a persistent depolarization). Readmission of oxygen and glucose immediately after the onset of rapid depolarization cannot restore the membrane potential. We have suggested that the rapid depolarization may result from a nonselective permeability to Na^+, Ca^{2+}, K^+, and Cl^-. Pretreatment of the slice with DL–2-amino-5-phosphonopentanoic acid (AP5), 6-cyano-7-nitroquinoxaline-2,3-dione (CNQX), or *trans*-1-aminocyclo-pentane-1, 3-dicarboxylate (*t*-ACPD) cannot prevent the rapid depolarization but does block the persistent depolarization; the membrane is repolarized to the control level after readmission of oxygen and glucose. Although neither, L-2-amino-3-phosphonopropionate (L-AP3) nor L-2-amino-4-phosphono-butylate (L-AP4) affects the rapid depolarization, either can depress the persistent depolarization; the membrane potential is not recovered. *t*-ACPD has shortened the duration of the Ca^{2+}-dependent spike in normoxic media, but AP5, CNQX, L-AP3, or L-AP4 does not. The protective effect of *t*-ACPD on the cell membrane is blocked by L-AP3 but L-AP3 cannot antagonize the shortening of the duration of the Ca^{2+}-dependent spike by *t*-ACPD. Therefore, Ca^{2+} influx via both the *N*-methyl-D-aspartate (NMDA) receptor and α-amino-3-hydroxy-5-methyl-4-isoxazolepropionate (AMPA)/kainate type II receptor contributes to the membrane dysfunction, and *t*-ACPD exerts protective action through an L-AP3-sensitive mechanism. It is highly possible that shortening of the Ca^{2+}-dependent spike by *t*-ACPD is caused by the activation of the metabotropic glutamate receptor 2 or 3. In a hippocampal slice loaded with fura-2-pentaacetoxymethyl ester, the intracellular concentration of Ca^{2+} gradually increases after 2 min of ischemic exposure and markedly elevates after approximately 5 min of exposure. The onset of the marked elevation of the intracellular Ca^{2+} concentration is consistent with the onset of rapid depolarization. All findings support the concept that activation of the NMDA receptor channel and the AMPA/kainate type II receptor channel has a critical role for membrane dysfunction caused by ischemic exposure.

Department of Physiology, Kurume University School of Medicine, 67 Asahi-machi, Kurume 830, Japan

Kew words. Ischemia—Oxygen and glucose deprivation—Intracellular Ca^{2+} concentration—NMDA receptors—Metabotropic glutamate receptors—Hippocampal slices

Introduction

Sensitivity to a transient or persistent ischemic insult varies considerably among central nervous system (CNS) neurons. Application of severe ischemia for only a few minutes can produce irreversible damage to neurons in the CA1 and CA4 region of the hippocampus, in the striatum, in layers III, V, and VI of the neocortex, and in the crebellar Purkinje cell layer [1–4]. A number of studies have focused on the metabolic energy deficit caused by the gap between oxygen demand and supply. It is well established that cerebral anoxia or ischemia decreases cellular energy production [5], leading to dysfunction of the Na^+ and Ca^{2+} pumps, which in turn results in marked changes of intracellular and extracellular ion concentrations [6]. It has been shown that an increase of extracellular Mg^{2+} ($[Mg^{2+}]_o$) or a decrease of extracellular Ca^{2+} ($[Ca^{2+}]_o$) protects hippocampal neurons in vitro from ischemic damage, and this effect seems to be the result of the block of synaptic transmission [7,8].

Excitatory amino acids, in addition to their role in neurotransmission and synaptic plasticity, have been implicated in neurotoxicity, as originally proposed by Olney et al. in the "excitotoxicity hypothesis" [9]. Hippocampal CA1 neurons receive their major excitatory inputs from pathways that release glutamate (Glu) or aspartate [10–12]. The excitotoxic hypothesis received support from the finding that the extracellular Glu concentration ($[Glu]_o$) in the hippocampus increases during hypoxia and ischemia [13–16], as well as from the finding that ischemic damage in the CA1 region can be reduced by collapse of the excitatory pathways to CA1 pyramidal cells [17–20] or by administration of *N*-methyl-D-aspartate (NMDA) receptor antagonists [21–23]. Moreover, NMDA receptor antagonists provide protection even when administered during reperfusion after occlusion of the common carotid arteries [22,24].

Studies on cultured cortical neurons [25] and cerebellar slices [26] have shown that cell death produced by activation of the NMDA receptor is caused primarily by excessive influx of Ca^{2+} through the receptor-operated channels [27,28], leading to an increase in intracellular Ca^{2+} concentration ($[Ca^{2+}]_i$) [29–32]. An elevated cytosolic Ca^{2+} may activate degradative enzymes, such as Ca^{2+}-activated phospholipases and neutral proteases [33]. Furthermore, delayed neuronal death is distinguishable from acute ischemic cell change by microscopic observation following ischemia. Delayed neuronal death occurs after 2 days of reperfusion from 5 min of ischemia [34].

The aim of this chapter is to discuss the contribution of the activation of non-NMDA, NMDA, or metabotropic glutamate (mGlu) receptors to the irreversible changes induced by acute ischemia in hippocampal CA1 and other CNS neurons.

Ischemia-Induced Potential Changes

In response to ischemia, the DC potential of the rat brain cortex in situ shows a transient, positive-going shift followed by a rapid negative shift (the so-called anoxic depolarization [35]) that is maintained. In response to anoxia, the DC potential shows a similar change. After reoxygenation, the DC potential, however, returns to the control level.

In CA1 neurons of a hippocampal slice preparation, superfusion with a medium equilibrated with 95% N_2–5% CO_2 (hypoxic medium) produces a transient membrane hyperpolarization followed by a slow depolarization, which within 20 min from the beginning of hypoxic exposure reaches a plateau of about 25 mV above the control resting potential. Both initial hyperpolarization and subsequent depolarization are associated with a decrease in membrane resistance. After reoxygenation, the membrane immediately begins to repolarize and shows a transient hyperpolarization. The membrane potential and input resistance return to prehypoxic levels after 15–20 min of reoxygenation [36] (see also [37–39] for early hyperpolarization or anoxic hyperpolarization).

In hippocampal CA1 neurons in vitro, superfusion with a hypoxic medium deprived of glucose (an ischemia-simulating medium) produces a transient initial hyperpolarization after 1.5 min of exposure to the medium, followed by a slow depolarization after 4–5 min and a rapid depolarization after approximately 6 min [40–42]. When oxygen and glucose are readmitted after the rapid depolarization has reached a value of −15 mV, the membrane potential not only does not repolarize, but depolarizes further, reaching 0 mV 4–5 min after readmission; thus, the neuron shows no functional recovery (also see [43]). In contrast with the hypoxic medium, which causes degeneration in only 21% of the recorded hippocampal CA1 neurons [36], the ischemic medium causes degeneration after approximately 6 min of superfusion in all hippocampal CA1 neurons recorded. The changes in membrane potential produced by the ischemic medium are a mirror image of the DC potential changes produced by ischemia or asphysia in the rat brain cortex [44], suggesting that this in vitro experimental system reflects the events occurring in the whole brain in situ.

Initial Hyperpolarization

There is general agreement that the initial hyperpolarization is generated by an increase in K^+ conductance [36,38,39]. The mechanism of the increase in K^+ conductance is not clear. It has been suggested that some of the early increase in $[Ca^{2+}]_i$ induced by the hypoxic medium is the result of Ca^{2+} mobilization from intracellular stores [45,46]: this could lead to an increase in Ca^{2+}-dependent K^+ conductance ($G_{K(Ca)}$). Direct supporting evidence, however, is either lacking [36] or is not conclusive [39]. Higashi et al. [47] have reported that the initial hyperpolarization is markedly depressed by a decrease in $[Ca^{2+}]_o$ or by addition of 1,2-bis(2-aminophenoxy)ethane-*N,N,N′,N′*-tetra-acetic acid (BAPTA/AM), methoxyverapamil (D-600), ryanodine, or procaine, but is not significantly affected by 4-aminopyridine or tetraethylammonium. Trifluoperazine or *N*-(6-aminohexyl)-5-chloro-1-naphthalenesulfonamide (W-7) suppresses the hyperpolarization. These findings indicate that hyperpolarization is caused by an increase in $G_{K(Ca)}$ that may be mediated intracellularly by calmodulin. Hyperpolarization is depressed by forskolin, N^6, 2′-*O*-dibutyryladenosine-3′:5′-cyclic monophosphate, *d*-tubocurarine, Ba^{2+}, and Cs^+. It is likely, therefore, that the pharmacological characteristics of the $G_{K(Ca)}$ activated by hypoxia are similar, but not identical, to those of the voltage-dependent $G_{K(Ca)}$ found in hippocampal and other CNS neurons. Another possibility is that a specific K^+ conductance is increased by a fall in cellular adenosine 5′-triphosphate ($G_{K(ATP)}$). Systematic tests with tolbutamide, a selective $G_{K(ATP)}$ antagonist, have not shown any significant reduction of the hyperpolarizing conductance increase [39,48].

Slow Depolarization

Superfusion of normoxic ouabain-containing medium mimics the initial hyperpolarization and subsequent slow depolarization induced by hypoxic exposure. The slow depolarization is also mimicked by elevation of the extracellular K^+ concentration ($[K^+]_o$) to 10–20 m*M*. These results suggest that depression of the electrogenic Na^+ pump activity and the resultant elevation of $[K^+]_o$ are involved in slow depolarization. Because the slow depolarization is not significantly altered by either low-Na^+ or low-Cl^- medium or Co^{2+}-containing solution, contribution of ions other than K^+ appears to be unlikely [36] (also see [49]).

Ischemic Rapid and Persistent Depolarizations

Tanaka et al. [50] have reported that the amplitude, maximal slope, and reversal potential of the rapid depolarization induced by the ischemic medium are 48 mV, 8.8 mV/s, and −13 mV, respectively. Reduction in extracellular Na^+ concentration ($[Na^+]_o$ from 143 m*M* to 14.3 m*M*) or $[Ca^{2+}]_o$ (from 2.5 m*M* to 0.25 m*M*) decreases the maximal slope and amplitude of the rapid depolarization and shifts the reversal potential in the hyperpolarizing direction. Reduction in extracellular Cl^- concentration ($[Cl^-]_o$ from 128 m*M* to 43 m*M*) increases the amplitude and shifts the reversal potential in the depolarizing direction without affecting the maximal slope. Reduction in $[K^+]_o$ (from 3.6 m*M* to 0.36 m*M*) does not change the amplitude and reversal potential, although it slightly decreases the maximal slope.

These results suggest that the rapid depolarization is a Ca^{2+}-, Na^+-, and Cl^--dependent potential. The lack of effect of the low-K^+ medium on rapid depolarization probably results because the K^+ concentration in the neuron environment is also determined by the increased K^+ efflux consequent to depression of the electrogenic Na^+ pump and to the increased $G_{K(Ca)}$ already discussed in the section on the initial hyperpolarization. In this regard, it is well known that, associated with the rapid negative shift of the DC potential of the rat neocortex in situ, $[K^+]_o$ increases while $[Ca^{2+}]_o$, $[Na^+]_o$, $[Cl^-]_o$, and extracellular volume decrease abruptly [6], as described next.

Rader and Lanthorn [43] have reported that the rapid depolarization recorded in single neurons intracellularly corresponds to the extracellularly recorded anoxic depolarization (rapid negative shift of the DC potential). This intracellularly recorded depolarization consists of two pharmacologically distinct components, a rapid depolarization and a subsequent, persitent depolarization. The persistent depolarization is selectively blocked by reduction in $[Ca^{2+}]_o$ or elevation of $[Mg^{2+}]_o$ or by NMDA receptor antagonists.

Alterations of Extracellular and Intracellular Ion Concentrations

The alterations of extracellular ion concentrations produced by ischemia were reported by Hansen and Zeuthen [51] in rat brain cortex in situ. Initially, during the positive-going shift of the DC potential, an increase of $[K^+]_o$ from 3 m*M* to approximately 10 m*M* is observed, while $[Ca^{2+}]_o$, $[Na^+]_o$, and $[Cl^-]_o$ show only minor changes. When the DC potential shows a rapid negative-going shift, $[K^+]_o$ exhibits a rapid increase from 10 m*M* to 55 m*M* while $[Na^+]_o$ rapidly decreases from 154 m*M* to 60-

mM, $[Cl^-]_o$ decreases from 129 mM to 75 mM, and $[Ca^{2+}]_o$ decreases from 1.3 mM to 80 nM. These changes in interstitial ion concentrations most likely result from efflux of K^+ from, and influx of Na^+, Cl^-, and Ca^{2+} into, neurons and glia cells [52,53]. In rat hypoglossal neurons in vitro, intracelluair Cl^- concentration ($[Cl^-]_i$), measured before and during anoxia, was 11 mM before anoxia and 21 mM during the slow depolarization after 3 min of anoxia [54]. In addition, intracellular pH ($[pH]_i$) decreased from 7.1 to 6.9 after 5 min of superfusion of rat hippocampal slices with O_2-and glucose-free sloution [49]. Hypoxia also decreases extracellular pH ($[pH]_o$) from 7.1 to 6.8 in the slice preparation of dorsal vagal motoneurons [55].

Silver and Erecinska [30] reported that in hippocampal CA1 neurons in vivo, the $[Ca^{2+}]_i$ is in the range of 10–100 nM under normoxia and increases to 0.6–0.8 mM after 2–4 min of ischemia, when there is also a rapid increase of $[K^+]_o$. In hippocampal slices loaded with Fura-2/AM, superfusion with the ischemic medium causes a gradual elevation of $[Ca^{2+}]_i$ within 5 min and a rapid, marked elevation of $[Ca^{2+}]_i$ after 5–10 min of exposure. The onset of the rapid $[Ca^{2+}]_i$ elevation is consistent with the onset of the rapid membrane depolarization [40]. Other in vitro studies also demonstrate accumulation of $[Ca^{2+}]_i$ induced by the hypoxic medium in acute dissociated rat CA1 neurons, using Ca^{2+}-sensitive fluorescence probes [56,57].

There are five possible routes for Ca^{2+} moving into the cytosol during anoxia or ischemia: voltage-gated Ca^{2+} channels, Ca^{2+} mobilization from intracellular storage sites, reverse operation of the Na^+/Ca^{2+} exchanger, leakage through damaged cell membrane, and transmitter-gated channels [58].

Dihydropyridine analogs such as nifedipine and nimodipine are potent L-type Ca^{2+} channel blockers. There is a report that nimodipine attenuates the increase in cytosolic $[Ca^{2+}]$ and neuronal damage caused by ischemia in vivo [59], but another study has shown that nimodipine fails to prevent ischemic neuronal damage in vivo [60]. A fluorescent imaging study in dissociated hippocampal CA1 neurons suggested that Ca^{2+} enters through Co^{2+}-sensitive Ca^{2+} channels during anoxia [57]. On the other hand, a voltage-clamp study suggested that in the CA1 neurons in vitro, the voltage-dependent (L-type and probably T-type, but not N-type) Ca^{2+} currents are suppressed by anoxia [61]. A selective N-type Ca^{2+} channel blocker, synthetic ω-conotoxin peptide SNX-111, protects hippocampal CA1 pyramidal neurons from the damage caused by 15 min of forebrain ischemia [62]. In rat hippocampal CA1 neurons, tetrodotoxin (TTX) reduces the neuronal damage caused by 20 min of ischemia in a limited manner but dose dependently [63]. It has been reported that voltage-dependent, TTX-sensitive Ca^{2+} channels exist in isolated CA1 hippocampal neurons that are dominant in the temporal hippocampus [64,65]. This area of the hippocampus is most vulnerable to ischemic attack, suggesting that activation of TTX-sensitive Ca^{2+} channels may be involved in the neuronal damage.

A number of studies have suggested that a massive Glu release induced by ischemia increases $[Ca^{2+}]_i$ as a result of activation of NMDA receptor channels [66–68]. In fact, in cultured hippocampal neurons, exogenous Glu markedly increases $[Ca^{2+}]_i$, and the increase is blocked by Mg^{2+}, a NMDA receptor channel blocker [69]. Among Glu agonists such as Glu, NMDA, quisqualate, kainate, α-amino-3-hydroxy-5-methyl-4-isoxazolepropionate (AMPA), and *trans*-1-aminocyclo-pentane-1,3-dicarboxylate (*t*-ACPD), NMDA is the most effective in increasing $[Ca^{2+}]_i$ in the stratum radiatum in the CA1 region of the hippocampal slice [70]. Murphy et al. [71] have shown that DL-2-amino-5-phosphonopentanoic acid (AP5), a competitive NMDA receptor antagonist, antagonizes the rise in $[Ca^{2+}]_i$ induced by NMDA in striatal cultured neurons.

It has been shown that intracellular Ca^{2+} mobilization induced by NMDA is potentiated by the initial Ca^{2+} influx into the cell and that intracellular Ca^{2+} mobilization is depressed by a blocker of adenosine diphosphate (ADP) ribosylation of G protein (G_i), an islet-activating protein [72]. The increased $[Ca^{2+}]_i$ may cause irreversible membrane damage that results from activation of Ca^{2+}-dependent enzymes in the cytosol. Staurosporine, a protein kinase C inhibitor, protects the rat and gerbil hippocampus from ischemic damage [73]. In hippocampal slices, a Ca^{2+}-dependent protease inhibitor, leupeptin, depresses accumulation of spectrin breakdown products in vulnerable neurons after ischemia [74]. NMDA receptor activation stimulates production of superoxide, which is correlated to the NMDA-induced neurotoxicity in mouse cerebellar granule cell culture [75]. Moreover, activation of nitric oxide synthase may be involved in the NMDA receptor-induced neurotoxicity in rat cortex, striatum, and hippocampus cell culture [76].

In Ca^{2+}-free and ethyleneglycoltetraacetic acid-(EGTA-) (0.5 m*M*) containing medium, however, oxygen and glucose deprivation can produce a marked increase in $[Ca^{2+}]_i$ in hippocampal tissue slice preparations [77]. Thus, the main source of increased $[Ca^{2+}]_i$ seems to be the Ca^{2+} release from intracellular storage sites. The mechanism of the intracellular Ca^{2+} release is still unclear. One possibility is that the ischemic condition causes elevation of $[Na^+]_i$ and $[H^+]_i$, which induces Ca^{2+} efflux from the store sites via the Na^+/Ca^{2+} exchanger and Ca^{2+}/H^+ antiporter, respectively. Other possibilities are a Ca^{2+}-activated Ca^{2+} release, i.e., the gradual elevation of $[Ca^{2+}]_i$ and $[Na^+]_i$ at the early stage of ischemic exposure triggers a massive Ca^{2+} release from the store site, a Na^+-dependent Ca^{2+} efflux from mitochondria, and an inositol triphosphate- (IP_3-) induced Ca^{2+} release from the intracellular storage site (see *Effects of Metabotropic Glutamate Receptor Agonists and Antagonist*, this chapter).

Glutamate Accumulation in the Interstitial Space

An in situ experiment has demonstrated a sevenfold increase in the concentration of Glu in the interstitial space ($[Glu]_o$) (from 138 μ*M* to 968 μ*M*) by ischemia [16]. $[Glu]_o$ increases gradually during the ischemic insult and increases more markedly after the end of the ischemia. Other studies have shown that the concentration of aspartate, glycine, γ-aminobutyric acid (GABA), dopamine, and serotonin in the interstitial space increases during ischemia [78,79].

The elevated $[Glu]_o$ during ischemia may be attributable to an increase in release and/or a decrease of uptake. Fifteen percent of the total [Glu] of the synaptosome of cerebral cortex or hippocampus is released in a Ca^{2+}-dependent manner, while the remaining 85% is released in a Ca^{2+}-independent manner [80]. Co^{2+} inhibits the increase in $[Glu]_o$ induced by the ischemic condition in vivo and in vitro [81]. Moreover, anoxia reduces the Ca^{2+}-dependent Glu release [82,83] and increases the Ca^{2+}-independent Glu release produced by reversing the Glu uptake system in vitro [80,84]. An in vivo study has shown biphasic increases in $[Glu]_o$ during the ischemia, i.e., an early increase, concomitant with the massive increase of $[K^+]_o$, is probably accompanied by rapid depolarization and followed by a subsequent, slow increase [85]. The early increase is completely blocked by Ca^{2+}-free and by Co^{2+}-containing solutions, while the later, slow increase is only partially blocked. From these results it has been suggested that the early increase, concomitant with the rapid $[K^+]_o$ increase, is caused mainly by Ca^{2+}-dependent Glu release, while the later increase results from both Ca^{2+}-dependent and Ca^{2+}-independent releases.

Alterations of Synaptic Responses During Ischemia

Depression of evoked excitatory postsynaptic potentials (EPSPs) caused by oxygen and/or glucose deprivation has been reported in many CNS neurons. In the stratum radiatum of the hippocampal CA1 region, the population spike is abolished and the remaining field EPSPs are depressed during anoxia [38]. In hippocampal CA1 pyramidal cells, the evoked EPSPs are well preserved, but the evoked inhibitory postsynaptic potentials (IPSPs) are completely abolished after 4 min of exposure to hypoxia [36]. A number of studies have confirmed the vulnerability of IPSPs to anoxia in hippocampal slices [86] and neocortical slices [87,88]. In contrast, it has been shown that the directly activated IPSPs or inhibitory postsynaptic currents (IPSCs), which are mediated by $GABA_A$ receptors, are resistant to anoxia while excitatory postsynaptic currents (EPSCs) are abolished [89,90]. These results indicate that the monosynaptic GABAergic neurotransmission is resistant to anoxia but the polysynaptic, GABAergic neurotransmission is relatively vulnerable.

The mechanisms underlying depressions of the evoked EPSP (EPSC) and IPSP (IPSC) are still not clarified. Because input resistance during the initial hyperpolarization is decreased to 40%–50% of the control value, suppression of both the EPSP and IPSP can be partially explained by the shunting effect on the synaptic potentials. Other possible explanations are, for example: (1) depression of spike generation in CA3 pyramidal neurons and basket cells, (2) blocking effects on impulse invasion into the presynaptic nerve terminals, (3) reduction of transmitter release from the presynaptic sites, (4) depression of sensitivity of the receptors, (5) blocking actions on the channels linked with the receptors, and (6) changes in the intracellular signal transduction mechanism.

In hippocampal pyramidal neurons, the increase in Cl^- conductance induced by GABA is well maintained after 25 min of hypoxic exposure [38]. Extracellular field recordings from the stratum radiatum, where excitatory afferent fibers make synapse with CA1 neurons, have shown that the presynaptic volley was robust during hypoxic exposure [91]. Repetitive orthodromic activation of Schaffer collateral caused a transient decrease in $[Ca^{2+}]_o$ measured in the stratum radiatum, which is attributable to influx of Ca^{2+} in presynaptic terminals. The transient decrease in $[Ca^{2+}]_o$ is rapidly and reversibly suppressed when oxygen is withdrawn from hippocampal tissue slices [92]. From these results Young and Somjen [92] concluded that presynaptic Ca^{2+} currents fail during hypoxia, perhaps because "resting" $[Ca^{2+}]_i$ is increased and, in part, also because of partial failure of presynaptic impulse conduction.

When energy was deprived by oxygen and glucose withdrawal, $[adenosine]_o$ increased from $14\,ng\,g^{-1}\,min^{-1}$ to $440\,ng\,g^{-1}\,min^{-1}$, most likely because of breakdown of intracellular ATP to adenosine 5′-diphosphate and adenosine 5′-monophosophate [93]. It has been shown that adenosine inhibits spontaneous and evoked EPSPs but not IPSPs in hippocampus slices [94,95]. Moreover, adenosine decreases K^+-stimulated Glu release and inhibits N-type Ca^{2+} channels in the CA3 pyramidal cell body [96] and nerve terminal [12]. Adenosine receptor antagonists attenuate the hypoxia-induced decrease of the evoked population spikes and EPSPs in hippocampal slices [97,98]. These results suggest that depression of evoked EPSP during anoxia or hypoxia may be the result of depression by adenosine of Glu release from the presynaptic nerve terminal. On the other hand, low doses of adenosine increase the evoked EPSPs in hippocampal CA3 neurons [99]. Thus, further work is necessary to elucidate the mechanisms involved.

In the CA1 region of guinea pig hippocampal slices, the population spike and EPSP gradually decreased during 10 min of superfusion with hypoxic medium, while $[Glu]_o$ did not change during the superfusion, but increased at the end of superfusion [100]. It is therefore unlikely that the depression of EPSPs at the early stage of hypoxia results from the small $[Glu]_o$ increase. In the rat hippocampal CA1 region in situ, $[Glu]_o$ is markedly increased and the evoked population EPSPs are abolished when the massive increase of $[K^+]_o$ occurs after 2 min of anoxia [85]. Superfusion with glucose-deficient medium inhibits excitatory synaptic transmission and suppresses the inhibitory effect of adenosine on K^+-evoked Glu release [101]. This result is consistent with the finding that the Ca^{2+}-independent Glu release is caused by reverse operation of Glu uptake after a few minutes of exposure to energy deprivation [102].

Action of Glutamate Receptor Antagonists on Ischemic Damage

Glu receptors can be categorized into two groups, ionotropic and metabotropic, on the basis of electrophysiological, pharmacological, and biochemical properties. Ionotropic receptors consist of NMDA and non-NMDA (AMPA/kainate) receptors. The ionotropic receptors contain cation-specific ion channels. Na^+, K^+, and Ca^{2+} pass through the NMDA receptor channel and type II AMPA/kainate receptor channel [28,103], and Na^+ and K^+ are permeable through the type I AMPA/kainate receptor channel [103,104].

A number of pathological studies have shown that the NMDA receptor antagonists 2-amino-7-phosphonoheptanoic acid (AP7), 3-([+]-2-carboxy-piperazin-4-yl)-propyl-1-phosphonic acid (CPP), *cis*-4-phosphonomethyl-2-piperidine carboxylate (CGS19755), and dizocilpine malate (MK-801) protect against the acute neuronal loss observed in rat hippocampus 2 h after recirculation from 30 min of ischemia [105] and the delayed neuronal loss observed in the hippocampus 7 days after 10 min of ischemia [106–109]. By contrast, the NMDA receptor antagonists ketamine and MK-801 do not have a significant protective action against neuronal loss or damage caused by 5–15 min of ischemia in rat hippocampus [110,111] or by 6 h of occlusion of one middle cerebral artery in the hippocampus [112]. Thus there is some disagreement about the protective action of NMDA receptor antagonists against ischemic damage in vivo.

2,3-Dihydroxy-6-nitro-7-sulfamoyl-benzo(F)quinoxaline (NBQX) or LY-293558, an AMPA/kainate type receptor antagonist, protects gerbil hippocampal CA1 neurons from selective neuronal loss for 4–5 days after 5 min of ischemia [113–115]. A 6-cyano-7-nitroquinoxaline-2,3-dione (CNQX), another AMPA/kainate receptor antagonist, protects against ischemic degeneration of retinal cells in vivo [116]. Therefore, it seems that AMPA/kainate receptor antagonists protect against ischemic neuronal degeneration in vivo.

In rat hippocampal CA1 neurons in vitro, a reduction in $[Ca^{2+}]_o$ and rise in $[Mg^{2+}]_o$ or application of CPP or MK-801 selectively block the persistent depolarization induced by hypoxia and low-D-glucose medium. As a result, the membrane potential is restored to the pre-hypoxic level after superfusing oxygenated artificial cerebrospinal fluid (ACSF) [43]. NMDA receptor antagonists restore the field EPSPs depressed by 15 min of superfusion of the ischemic medium in rat olfactory cortex slices [117]. Thus, NMDA receptor antagonists protect against ischemic neuronal damage in vitro.

A quisqualate-induced neurotoxicity is attenuated by CNQX in mouse cultured cells [118]. On the other hand, in the hippocampal slice preparation, quisqualate neurotoxicity results from activation of CNQX-resistant quisqualate receptors, suggesting that the neurotoxicity probably results from activation of metabotropic Glu receptor (mGluR) 1 or mGluR5 [119]. 5,7-Dinitroquinoxaline-2,3-dione (DNQX), an AMPA/kainate type receptor antagonist, fails to block the suppression of the field EPSPs caused by the ischemic medium.

Yamamoto et al. (Yamamoto S, Tanaka E, Shoji Y, Polosa C, Kudo Y, Higashi H: Effects of glutamate agonists and antagonists on the potential change induced by deprivation of oxygen and glucose in rat CA1 hippocampal neurons in vitro, in manuscript) have shown that AP5 and CNQX do not block the rapid depolarization induced by ischemic medium, but do block the subsequent, persistent depolarization in a majority of hippocampal CA1 neurons. These results indicate that activation of neither the NMDA nor the AMPA/kainate receptor contributes to the generation of the rapid depolarization. The activation of NMDA and AMPA/kainate receptors contributes to the generation of the persistent depolarization. AP5 and CNQX recovered the membrane potential, dose dependently. Reduction in $[Ca^{2+}]_o$ or addition of Co^{2+} or Mg^{2+} does not prevent rapid depolarization, but blocks persistent depolarization, suggesting that rapid depolarization is not attributable to activation of voltage-dependent Ca^{2+} channels while persistent depolarization involves Ca^{2+} influx through NMDA receptors and AMPA/kainate type II receptors (Yamamoto et al., in manuscript).

Effects of Metabotropic Glutamate Receptor Agonists and Antagonists

mGluRs are divided into at least seven subtypes (mGluR1–mGluR7). When mGluR1 and mGluR5 are activated by quisqualate, they stimulate IP_3 formation and intracellular Ca^{2+} mobilization. Activation of mGluR2 and mGluR3 by L-Glu or *t*-ACPD markedly decreases intracellular adenosine 3′,5′-monophosphate (cAMP). 2-Amino-4-phosphonobutylate (AP4) activates mGluR4, which decreases cAMP concentration and depresses Glu-mediated EPSC by a presynaptic mechanism in cultured hippocampal neurons [120]. Baskys and Malenka [121] reported that *t*-ACPD, ibotenate, AP4, and quisqualate suppress presynaptically excitatory synaptic transmission in rat hippocampal CA1 slices. Therefore, mGluR4 (an AP4-sensitive receptor), mGluR2, and mGluR3 are so-called autoreceptors [104]. The roles of mGluR6 and mGluR7 are still unclear [122,123].

Garthwaite and Garthwaite [119] inferred an ischemic damage induced by activation of mGluRs from the finding of CNQX-insensitive quisqualate neurotoxicity in hippocampal slices. In contrast, the metabotropic receptor agonist *t*-CAPD attenuates NMDA neurotoxicity [124], and by itself does not mediate neuronal degeneration in cortical cultured cells [125]. Moreover, neuroprotection by *t*-ACPD is partially reversed by the protein kinase antagonists 1-(5-isoquinolinesulfonyl)-2-methylpiperazine (H-7) and *N*-(2-guanidino-ethyl)-5-isoquinolinesulfonamide (HA-

1004), suggesting that a concomitant activation of protein kinases by *t*-ACPD may serve as a protective mechanism against NMDA neurotoxicity [124]. Similarly, administration of *t*-ACPD reduced the infarct volume in mice after middle cerebral artery occlusion [126]. In rat hippocampal slices, the field EPSP of the CA1 region showed a significant recovery in the presence of *t*-ACPD during reoxygenation after hypoxic exposure [127].

Yamamoto et al. (in manuscript) have shown that *t*-ACPD at higher concentration ($>20\,\mu M$) does not prevent rapid depolarization induced by the ischemic medium, but blocks the subsequent persistent depolarization; the membrane potential is repolarized to the resting membrane potential after reintroducing oxygen and glucose. AP4 (1 m*M*) does not affect rapid depolarization but depresses persistent depolarization. However, AP4 has not been shown to restore the depolarized membrane potential to resting membrane potential after reintroducing oxygen and glucose in most neurons. *t*-ACPD ($20\,\mu M$) reduces the duration of Ca^{2+}-dependent spikes in normoxic media but AP4 (1 m*M*) does not. Because AP4 (1 m*M*) markedly antagonizes ibotenate-induced phosphoinositide (PI) hydrolysis in rat hippocampal tissue slices [128,129], this suggests that activation of mGluR4, mGluR6, or mGluR7 does not contribute to the membrane dysfunction.

L-2-Amino-3-phosphonopropionate (L-AP3) is a metabotropic Glu receptor antagonist that is selective for mGluR1 and mGluR5 [129]. Opitz and Reymann [130] have shown that in the presence of L-AP3, hippocampal field EPSPs and population spikes are suppressed by the anoxic medium but recover to 80%–100% of control amplitudes after reoxygenation. In contrast, Yamamoto et al. (in manuscript) have shown that the protective effect of *t*-ACPD against the cell membrane dysfunction produced by oxygen-glucose deprivation is antagonized by L-AP3 (1 m*M*). L-AP3 itself only reduces persistent depolarization, antagonizing neither Ca^{2+}-dependent spike shortening by *t*-ACPD in normoxic media nor ischemic membrane dysfunction. These results indicate that *t*-ACPD exerts protective action through an L-AP3-sensitive mechanism. Because AP3 (1 m*M*) completely antagonizes ibotenate-induced PI hydrolysis in rat hippocampal tissue slices [129], this suggests that activation of mGluR1 or mGluR5 does not contribute to the membrane dysfunction.

t-ACPD stimulates mGluR1 and mGluR5, which cause PI hydrolysis, and increases $[Ca^{2+}]_i$. The effect is partially blocked by L-AP3 (1 m*M*) [131,132]. The $[Ca^{2+}]_i$ accumulation parallels the cytopathological changes both in time-course and in the site involved in NMDA neurotoxicity [133,134]. Thus, *t*-ACPD can in principle contribute to ischemic damage and "excitotoxicity," while L-AP3 cannot protect against neuronal degeneration. However, pretreatment with *t*-ACPD will reduce the Ca^{2+} concentration in intracellular storage sites because the $[Ca^{2+}]_i$ increased by *t*-ACPD returns to the normal concentration by action of the ATP-dependent Ca^{2+} pump and of the Na^+/Ca^{2+} exchanger in normoxic conditions.

The subtype of mGluRs involved in the neuroprotection of *t*-ACPD is, however, unresolved. AP4, an activator of presynaptic "autoreceptors," does not have any protective action against membrane dysfunction induced by the ischemic medium (Yamamoto et al., in manuscript), suggesting that the protective effect of *t*-ACPD does not result from activation of mGluR4, mGluR6, or mGluR7. Our preliminary finding, that *t*-ACPD partially protects at low concentrations ($<10\,\mu M$) and completely protects at high concentrations ($>20\,\mu M$) against the membrane dysfunction induced by the ischemic medium, suggests that the activation of mGluR2 and mGluR3 by *t*-ACPD would be at least in part attributed to the protective action, because mGluR2 and mGluR3 are activated by *t*-ACPD at low concentration (0.1–$10\,\mu M$) in oocytes [135].

However, data on the effect of L-AP3 on mGluR2 and mGluR3 are lacking (see following). L-AP3 (1 m*M*) decreases IP_3 formation induced through activation of mGluR1 (α) to 56% and to approximately 20% through mGluR1 (β) and mGluR5 activation, respectively, in oocyte and Chinese hamster ovary (CHO) cells [136–138]. If L-AP3 has similar inhibitory potencies on mGluR1 (α), mGluR1(β), and mGluR5 in hippocampal neurons, it is unlikely that the protective action of *t*-ACPD is involved in these mGluRs, because L-AP3 (1 m*M*) itself did not have any neuroprotection. Therefore, it is most likely that the protective action of *t*-ACPD may be caused by activation of mGluR1 or mGluR5. However, further work is necessary to rule out the possibility of the involvement of these mGluRs in *t*-ACPD neuroprotection.

In hippocampal cultured neurons, *t*-ACPD reversibly reduces high-threshold Ca^{2+} currents (both N-type and L-type) to 70% of the control, and L-AP3 blocks the *t*-ACPD-induced Ca^{2+} current reduction [139]. Quisqualate also depresses Ca^{2+} currents [140]. On the other hand, in hippocampal tissue slices, *t*-ACPD shortens the duration of the Ca^{2+}-dependent spike, but L-AP3 cannot antagonize spike shortening (Yamamoto et al., in manuscript). The result suggests that L-AP3-insensitive mGluRs (presumably mGluR1(β), mGluR2, mGluR3, and mGluR5) would be involved in the spike shortening. It is well known that stimulation of β-adrenergic receptors increases intracellular cAMP concentration ($[cAMP]_i$), which results in enhancement of Ca^{2+} currents in various neurons [141]; e.g., the current of single L-type channels in hippocampal and sympathetic neurons is augmented by cAMP [142,143]. Conversely, agents that inhibit cAMP formation reduce Ca^{2+} currents (see [141]). The activation of mGluR2 and mGluR3 reduces cAMP formation. Therefore, it is highly possible that Ca^{2+} spike shortening by *t*-ACPD results from the activation of mGluR2 and mGluR3.

Conclusion

In this chapter, we reviewed in vitro and in vivo studies of the excitatory amino acid neurotoxicity induced by acute, ischemic exposure and summarized the contribution of Glu receptor subtypes to neurotoxicity. The pharmacology of Glu antagonists in ischemic neurotoxicity is difficult to study in an in vivo preparation because even competitive antagonists have to be given in such high doses that the receptor selectivity becomes questionable. In contrast, under conditions in vitro where antagonist concentration and exposure time can be accurately controlled, non-NMDA (AMPA/kainate), NMDA, and metabotropic Glu toxicities target quite separate neuronal populations in brain tissue slices.

Hippocampal CA1 pyramidal cells are particularly vulnerable to brief episodes of ischemia or hypoxia in vivo [1]. In the hippocampal slice preparation, deprivation of both oxygen and glucose produces a transient hyperpolarization and a subsequent slow depolarization in CA1 pyramidal neurons. The slow depolarization is followed by a rapid depolarization after approximately 6 min of deprivation. Depolarization reaches −15 mV and the membrane potential slowly rises to 0 mV (a persistent depolarization). Readmission of oxygen and glucose immediately after the onset of rapid depolarization cannot restore the membrane potential [40–43]. The membrane potential change under an ischemic condition is a mirror image of the DC potential change by ischemia [44], e.g., rapid depolarization corresponds to a rapid negative shift of the DC potential [49].

Tanaka et al. [50] suggested that the rapid depolarization is caused by a nonselective permeability to Na^+, Ca^{2+}, K^+, and Cl^-. Pretreatment of the slice with AP5, CNQX, or *t*-ACPD cannot prevent rapid depolarization but blocks persistent depolarization; the membrane is repolarized to the control level after readmission of oxygen and glucose [48]; Yamamoto et al., (in manuscript). L-AP3 or L-AP4 did not affect rapid depolarization but depresses persistent depolarization. Therefore, neither NMDA receptor activation nor AMPA/kainate receptor activation contributes to the generation of rapid depolarization, while the Glu receptor activation mainly contributes to generation of the persistent depolarization. In other words, Glu accumulation is involved in the persistent depolarization.

In contrast to L-AP3 and L-AP4, CNQX and AP5 recovered membrane potential, suggesting that Ca^{2+} influx via both the AMPA/kainate type II receptor and the NMDA receptor contributes to membrane dysfunction. The protective action of *t*-ACPD is reversed by L-AP3, suggesting that the neuroprotection results from activation of AP3-sensitive mGluRs. However, the subtypes involved are still unknown. *t*-ACPD reduces the duration of the Ca^{2+}-dependent spike in normoxic media, and L-AP3 cannot reverse the spike shortening induced by *t*-ACPD (Yamamoto et al., in manuscript). The activation of β-adrenoceptors increases $[cAMP]_i$, which results in enhancement of Ca^{2+} currents in hippocampal neurons [141,142]. Conversely, agents that inhibit cAMP formation reduce Ca^{2+} currents (see [141]). The activation of mGluR2 and mGluR3 reduces cAMP formation. Therefore, it is highly possible that Ca^{2+} spike shortening by *t*-ACPD is caused by activation of mGluR2 and mGluR3.

In a hippocampal slice loaded with Fura-2/AM, $[Ca^{2+}]_i$ gradually increased after 2 min of ischemic exposure and markedly elevated after approximately 5 min of exposure. The onset of the marked $[Ca^{2+}]_i$ elevation is consistent with the onset of rapid depolarization [40]. Moreover, even in Ca^{2+}-free solution containing EGTA (0.5 m*M*), ischemic exposure markedly increases $[Ca^{2+}]_i$ [77]. Thus, the main source of increased $[Ca^{2+}]_i$ seems to be an intracellular Ca^{2+} release from an intracellular Ca^{2+} storage site. Although the increased $[Ca^{2+}]_i$ mainly results from intracellular Ca^{2+} release, all findings support the concept that activation of the NMDA receptor channel and the AMPA/kainate type II receptor channel has a critical role for membrane dysfunction caused by ischemic exposure.

References

1. Brierly JB, Graham DI (1984) Greenfield's neuropathology. Wiley, New York, pp 125–209
2. Jørgensen MB, Diemer NH (1982) Selective neuron loss after cerebral ischemia in the rat: possible role of transmitter glutamate. Acta Neurol Scand 66:536–546
3. Smith M-L, Auer RN, Siesjö BK (1984) The density and distribution of ischemic brain injury in the rat following 2–10 min of forebrain ischemia. Acta Neuropathol 64:319–332
4. Crain BJ, Westerkam WD, Harrison AH, Nadler JV (1988) Selective neuronal death after transient forebrain ischemia in the mongolian gerbil: a silver impregnation study. Neuroscience 27:387–402
5. Lowry OH, Passonneau JV, Hasselberger FX, Schulz DW (1964) Effect of ischemia on known substrates and cofactors of the glycolytic pathway in brain. J Biol Chem 239:18–30
6. Hansen AJ (1985) Effect of anoxia on ion distribution in the brain. Physiol Rev 65:101–148
7. Kass IS, Lipton P (1982) Mechanisms involved in irreversible anoxic damage to the in vitro rat hippocampal slice. J Physiol (Camb) 332:459–472

8. Rothman SM (1983) Synaptic activity mediates death of hypoxic neurons. Science 220:536–537
9. Olney JW, Ho OL, Rhee V (1971) Cytotoxic effects of acidic and sulphur containing amino acids on the infant mouse central nervous system. Exp Brain Res 14:61–76
10. Nadler JV, Vaca KW, White WF, Lynch GS, Cotman CW (1976) Aspartate and glutamate as possible transmitters of excitatory hippocampal afferents. Nature 260:538–540
11. Collingridge GL, Kehl SJ, McLennan H (1983) Excitatory amino acids in synaptic transmission in the schaffer collateral-commissural pathway of the rat hippocampus. J Physiol (Camb) 334:33–46
12. Burke SP, Nadler JV (1988) Regulation of glutamate and aspartate release from slices of the hippocampal CA1 area: effects of adenosine and baclofen. J Neurochem 51:1541–1551
13. Benveniste H, Drejer J, Schousboe A, Diemer NH (1984) Elevation of the extracellular concentrations of glutamate and aspartate in rat hippocampus during transient cerebral ischemia monitored by intracerebral microdialysis. J Neurochem 43:1369–1374
14. Hagberg H, Lehmann A, Sandberg M, Nyström B, Jacobson I, Hamberger A (1985) Ischemia-induced shift of inhibitory and excitatory amino acids from intra- to extracellular compartments. J Cereb Blood Flow Metab 5:413–419
15. Ikeda M, Nakazawa T, Abe K, Kaneko T, Yamatsu K (1989) Extracellular accumulation of glutamate in the hippocampus induced by ischemia is not calcium dependent—in vitro and in vivo evidence. Neurosci Lett 96:202–206
16. Benveniste H (1991) The excitotoxin hypothesis in relation to cerebral ischemia. Cerebrovasc Brain Metab Rev 3:213–245
17. Wieloch T, Lindval O, Blomqvist P, Gage FH (1985) Evidence for amelioration of ischemic neuronal damage in the hippocampal formation by lesion of the perforanth path. Neurol Res 7:24–26
18. Onodera H, Sato G, Kogure K (1986) Lesions to Schaffer collaterals prevent ischemic death of CA1 pyramidal cells. Neurosci Lett 68:169–174
19. Johansen FF, Jørgensen MB, Diemer NH (1986) Ischemic CA1 pyramidal cell loss is prevented by colchicine destruction of dentate gyrus granule cells. Brain Res 377:344–347
20. Jørgensen MB, Johansen FF, Diemer NH (1987) Removal of the enthorhinal cortex protects hippocampal CA1 neurons from ischemic damage. Acta Neuropathol 73:189–194
21. Boast CA, Gerhardt SC, Janak P (1987) Excitatory amino acid transmission. Liss, New York, pp 249–296
22. Boast CA, Gerhardt SC, Pastor G (1988) The *N*-methyl-D-aspartate antagonists CGS19755 and CPP reduce ischemic brain damage in gerbils. Brain Res 442:345–348
23. Gill R, Foster AC, Woodruff GN (1987) Systemic administration of MK-801 protects against ischemia-induced hippocampal neurodegeneration in the gerbil. J Neurosci 7:3343–3349
24. Gill R, Foster AC, Woodruff GN (1988) MK-801 is neuroprotective in gerbils when administered during the post-ischemic period. Neuroscience 25:847–855
25. Choi DW (1987) Ionic dependence of glutamate neurotoxicity. J Neurosci 7:369–379
26. Garthwaite G, Garthwaite J (1987) Receptor-linked ionic channels mediate *N*-methyl-D-aspartate neurotoxicity in rat cerebellar slices. Neurosci Lett 83:241–246
27. MacDermott AB, Mayer ML, Westbrook GL, Smith SJ, Barker JL (1986) NMDA-receptor activation increases cytoplasmic calcium concentration in cultured spinal cord neurones. Nature 321:519–522
28. Mayer ML, Westbrook GL (1987) Permeation and block of *N*-methyl-D-aspartic acid receptor channels by divalent cations in mouse cultured central neurones. J Physiol (Camb) 394:501–527
29. Garthwaite G, Garthwaite J (1989) Differential dependence on Ca^{2+} of *N*-methyl-D-aspartate and quisqualate neurotoxicity in young rat hippocampal slices. Neurosci Lett 97:316–322

30. Silver IA, Erecinska M (1990) Intracellular and extracellular changes of [Ca^{2+}] in hypoxia and ischemia in rat brain *in vivo*. J Gen Physiol 95:837–866
31. Michaels RL, Rothman SM (1990) Glutamate neurotoxicity *in vitro*: antagonist pharmacology and intracellular calcium concentrations. J Neurosci 10:283–292
32. Randall RD, Thayer SA (1992) Glutamate-induced calcium transient triggers delayed calcium overload and neurotoxicity in rat hippocampal neurons. J Neurosci 12:1882–1895
33. Orrenius S, McConkey DJ, Bellomo G, Nicotera P (1989) Role of Ca^{2+} in toxic cell killing. Trends Pharmacol Sci 10:281–285
34. Kirino T (1982) Delayed neuronal death in the gerbil hippocampus following ischemia. Brain Res 239:57–69
35. Hansen AJ (1977) Extracellular potassium concentration in juvenile and adult rat brain cortex during anoxia. Acta Physiol Scand 99:412–420
36. Fujiwara N, Higashi H, Shimoji K, Yoshimura M (1987) Effects of hypoxia on rat hippocampal neurones in vitro. J Physiol (Camb) 384:131–151
37. Donnelly DF, Jiang C, Haddad GG (1992) Comparative responses of brain stem and hippocampal neurons to O_2 deprivation: in vitro intracellular studies. Am J Physiol 262:L549–L554
38. Hansen AJ, Hounsgaard J, Jahnsen H (1982) Anoxia increases potassium conductance in hippocampal nerve cells. Acta Physiol Scand 115:301–310
39. Leblond J, Krnjević K (1989) Hypoxic changes in hippocampal neurons. J Neurophysiol (Bethesda) 62:1–14
40. Kudo Y, Akita K, Higashi H (1989) Temporal and spatial changes in intracellular Ca^{2+} concentration of rat hippocampal slice during hypoxia. Neurosci Res (Suppl) 9:S33
41. Higashi H (1990) Mechanisms of ischemic brain damage. PharMa International, Tokyo, pp 30–37
42. Higashi H, Kudo Y, Nishi S (1990) Dysfunction of rat hippocampal pyramidal cells induced by deprivation of oxygen and glucose. Neurosci Res (suppl) 11:S18
43. Rader RK, Lanthorn TH (1989) Experimental ischemia induces a persistent depolarization blocked by decreased calcium and NMDA antagonists. Neurosci Lett 99:125–130
44. Hansen AJ (1978) The extracellular potassium concentration in brain cortex following ischemia in hypo- and hyperglycemic rats. Acta Physiol Scand 102:324–329
45. Krnjević K, Xu YZ (1989) Dantrolene suppresses the hyperpolarization or outward current observed during anoxia in hippocampal neurons. Can J Physiol Pharmacol 67:1602–1604
46. Katchman AN, Hershkowitz N (1993) Early anoxia-induced vesicular glutamate release results from mobilization of calcium from intracellular stores. J Neurophysiol (Bethesda) 70:1–7
47. Higashi H, Yamamoto S, Nishi S (1992) Mechanisms for hypoxia-induced hyperpolarizations in the rat hippocampal CA1 neurons *in vitro*. Jpn J Physiol (Suppl) 42:S9
48. Krnjević K, Leblond J (1988) Are there hippocampal ATP-sensitive K-channels that are activated by anoxia? Pflüegers Arch (Suppl) 411:R145
49. Fujiwara N, Abe T, Endo H, Warashina A, Shimoji K (1992) Changes in intracellular pH of mouse hippocampal slices responding to hypoxia and/or glucose depletion. Brain Res 572:335–339
50. Tanaka E, Yamamoto S, Higashi H (1995) Ionic basis of the rapid depolarization induced by deprivation of oxygen and glucose in rat hippocampal CA1 neurons in vitro. Neurosci Res (Suppl) 19:S27
51. Hansen AJ, Zeuthen T (1981) Extracellular ion concentrations during spreading depression and ischemia in the rat brain cortex. Acta Physiol Scand 113:437–445
52. Rothman SM (1985) The neurotoxicity of excitatory amino acids is produced by passive chloride influx. J Neurosci 5:1483–1489
53. Siesjö BK (1988) Historical overview. Calcium, ischemia, and death of brain cells. Ann NY Acad Sci 522:638–661

54. Jiang C, Agulian S, Haddad GG (1992) Cl^- and Na^+ homeostasis during anoxia in rat hypoglossal neurons: intracellular and extracellular *in vitro* studies. J Physiol (Camb) 448:697–708
55. Martin RL, Lloyd HGE, Cowan AI (1994) The early events of oxvgen and glucose deprivation: setting the scene for neuronal death? Trends Neurosci 17:251–257
56. Mitani A, Kadoya F, Kataoka K (1990) Distribution of hypoxia-induced calcium accumulation in gerbil hippocampal slice. Neurosci Lett 120:42–45
57. Friedman JE, Haddad GG (1993) Major differences in $Ca^{2+}{}_i$ response to anoxia between neonatal and adult rat CA1 neurons: role of $Ca^{2+}{}_0$ and $Na^+{}_0$. J Neurosci 13:63–72
58. Choi DW (1988) Calcium-mediated neurotoxicity: relationship to specific channel types and role in ischemic damage. Trends Neurosci 11:465–469
59. Uematsu D, Greenberg JH, Hickey WF, Reivich M (1989) Nimodipine attenuates both increase in cytosolic free calcium and histologic damage following focal cerebral ischemia and reperfusion in cat. Stroke 20:1531–1537
60. Vibulsresth S, Dietrich WD, Busto R, Ginsberg MD (1987) Failure of nimodipine to prevent ischemic neuronal damage in rats. Stroke 18:210–216
61. Krnjević K, Leblond J (1989) Changes in membrane currents of hippocampal neurons evoked by brief anoxia. J Neurophysiol (Bethesda) 62:15–30
62. Valentino K, Newcomb R, Gadbois T, Singh T, Bowersox S, Bitner S, Justice A, Yamashiro D, Hoffman BB, Ciaranello R, Miljanich G, Ramachandran J (1993) A selective N-type calcium channel antagonist protects against neuronal loss after global cerebral ischemia. Proc Natl Acad Sci USA 90:7894–7897
63. Yamasaki Y, Kogure K, Hara H, Ban H, Akaike N (1991) The possible involvement of tetrodotoxin-sensitive ion channels in ischemic neuronal damage in the rat hippocampus. Neurosci Lett 121:251–254
64. Akaike N, Takahashi K, Morimoto M (1991) Heterogeneous distribution of tetrodotoxin-sensitive calcium-conducting channels in rat hippocampal CA1 neurons. Brain Res 556:135–138
65. Akaike N, Takahashi K (1992) Tetrodotoxin-sensitive calcium-conducting channels in the rat hippocampal CA1 region. J Physiol (Camb) 450:529–546
66. Vornov JJ, Coyle JT (1991) Enhancement of NMDA receptor-mediated neurotoxicity in the hippocampal slice by depolarization and ischemia. Brain Res 555:99–106
67. Salinska E, Pluta R, Puka M, Lazarewicz JW (1991) Blockade of *N*-methyl-D-aspartate-sensitive excitatory amino acid receptors with 2-amino-5-phosphonovalerate reduces ischemia-evoked calcium redistribution in rabbit hippocampus. Exp Neurol 112:89–94
68. Lobner D, Lipton P (1993) Intracellular calcium levels and calcium fluxes in the CA1 region of the rat hippocampal slice during in vitro ischemia: relationship to electrophysiological cell damage. J Neurosci 13:4861–4871
69. Kudo Y, Ogura A (1986) Glutamate-induced increase in intracellular Ca^{2+} concentration in isolated hippocampal neurones. Br J Pharmacol 89:191–198
70. Kudo Y, Nakamura T, Ito E (1991) A "macro" image analysis of fura-2 fluorescence to visualize the distribution of functional glutamate receptor subtypes in hippocampal slices. Neurosci Res 12:412–420
71. Murphy SN, Thayer SA, Miller RJ (1987) The effects of excitatory amino acids on intracellular calcium in single mouse striatal neurons in vitro. J Neurosci 7:4145–4158
72. Harada K, Yoshimura T, Nakajima K, Ito H, Ebina Y, Shingai R (1992) *N*-Methyl-D-aspartate increases cytosolic Ca^{2+} via G proteins in cultured hippocampal neurons. Am J Physiol 262:C870–C875
73. Hara H, Onodera H, Yoshidomi M, Matsuda Y, Kogure K (1990) Staurosporine, a novel protein kinase C inhibitor, prevents postischemic neuronal damage in the gerbil and rat. J Cereb Blood Flow Metab 10:646–653
74. Lee KS, Frank S, Vanderklish P, Arai A, Lynch G (1991) Inhibition of proteolysis protects hippocampal neurons from ischemia. Proc Natl Acad Sci USA 88:7233–7237

75. Lafon-Cazal M, Pietri S, Culcasi M, Bockaert J (1993) NMDA-dependent superoxide production and neurotoxicity. Nature 364:535–537
76. Dawson VL, Dawson TM, London ED, Bredt DS, Snyder SH (1991) Nitric oxide mediates glutamate neurotoxicity in primary cortical cultures. Proc Natl Acad Sci USA 88:6368–6371
77. Ebine Y, Fujiwara N, Shimoji K (1994) Mild acidosis inhibits the rise in intracellular Ca^{2+} concentration in response to oxygen-glucose deprivation in rat hippocampal slices. Neurosci Lett 168:155–158
78. Globus MY-T, Busto R, Martinez E, Valdés I, Dietrich WD, Ginsberg MD (1991) Comparative effect of transient global ischemia on extracellular levels of glutamate, glycine, and γ-aminobutyric acid in vulunerable and nonvulunerable brain regions in the rat. J Neurochem 57:470–478
79. Baker AJ, Zornow MH, Scheller MS, Yaksh TL, Skilling SR, Smullin DH, Larson AA, Kuczenski R (1991) Changes in extracellular concentrations of glutamate, aspartate, glycine, dopamine, serotonin, and dopamine metabolites after transient global ischemia in the rabbit brain. J Neurochem 57:1370–1379
80. Nicholls D, Attwell D (1990) The release and uptake of excitatory amino acids. Trends Pharmaco Sci 11:462–468
81. Drejer J, Benveniste H, Diemer NH, Schousboe A (1985) Cellular origin of ischemia-induced glutamate release from brain tissue in vivo and in vitro. J Neurochem 45:145–151
82. Kauppinen RA, McMahon HT, Nicholls DG (1988) Ca^{2+}-Dependent and Ca^{2+}-independent glutamate release, energy status and cytosolic free Ca^{2+} concentration in isolated nerve terminals following metabolic inhibition: possible relevance to hypoglycaemia and anoxia. Neuroscience 27:175–182
83. Sánchez-Prieto J, González P (1988) Occurrence of a large Ca^{2+}-independent release of glutamate during anoxia in isolated nerve terminals (synaptosomes). J Neurochem 50:1322–1324
84. Madl JE, Burgesser K (1993) Adenosine triphosphate depletion reverses sodium-dependent, neuronal uptake of glutamate in rat hippocampal slices. J Neurosci 13:4429–4444
85. Katayama Y, Kawamata T, Tamura T, Hovda DA, Becker DP, Tsubokawa T (1991) Calcium-dependent glutamate release concomitant with massive potassium flux during cerebral ischemia in vivo. Brain Res 558:136–140
86. Hershkowitz N, Katchman AN, Veregge S (1993) Site of synaptic depression during hypoxia: a patch-clamp analysis. J Neurophysiol (Bethesda) 69:432–441
87. Luhmann HJ, Heinemann U (1992) Hypoxia-induced functional alterations in adult rat neocortex. J Neurophysiol (Bethesda) 67:798–811
88. Rosen AS, Morris ME (1993) Anoxic depression of excitatory and inhibitory postsynaptic potentials in rat neocortical slices. J Neurophysiol (Bethesda) 69:109–117
89. Khazipov R, Bregestovski P, Ben-Ari Y (1993) Hippocampal inhibitory interneurons are functionally disconnected from excitatory inputs by anoxia. J Neurophysiol (Bethesda) 70:2251–2259
90. Zhu PJ, Krnjević K (1994) Anoxia selectively depresses excitatory synaptic transmission in hippocampal slices. Neurosci Lett 166:27–30
91. Taylor CP, Weber ML (1993) Effect of temperature on synaptic function after reduced oxygen and glucose in hippocampal slices. Neuroscience 52:555–562
92. Young JN, Somjen GG (1992) Suppression of presynaptic calcium currents by hypoxia in hippocampal tissue slices. Brain Res 573:70–76
93. Pedata F, Latini S, Pugliese AM, Pepeu G (1993) Investigations into the adenosine outflow from hippocampal slices evoked by ischemia-like conditions. J Neurochem 61:284–289
94. Schubert P, Mitzdorf U (1979) Analysis and quantitative evaluation of the depressive effect of adenosine on evoked potentials in hippocampal slices. Brain Res 172:186–190

95. Yoon K-W, Rothman SM (1991) Adenosine inhibits excitatory but not inhibitory synaptic transmission in the hippocampus. J Neurosci 11:1375–1380
96. Mogul DJ, Adams ME, Fox AP (1993) Differential activation of adenosine receptors decreases N-type but potentiates P-type Ca^{2+} current in hippocampal CA3 neurons. Neuron 10:327–334
97. Fowler JC (1990) Adenosine antagonists alter the synaptic response to *in vitro* ischemia in the rat hippocampus. Brain Res 509:331–334
98. Gribkoff VK, Bauman LA, Van der Maelen CP (1990) The adenosine antagonist 8-cyclopentyltheophylline reduces the depression of hippocampal neuronal responses during hypoxia. Brain Res 512:353–357
99. Sakurai T, Okada Y (1992) Excitatory effect of adenosine on neurotransmission and the release of glutamate from hippocampal slices of guinea-pig. J Physiol (Camb) 446:380P
100. Penning DH, Goh JW, El-Beheiry H, Brien JF (1993) Effect of hypoxia on glutamate efflux and synaptic transmission in the guinea pig hippocampus. Brain Res 620:301–304
101. Burke SP, Nadler JV (1989) Effects of glucose deficiency on glutamate/aspartate release and excitatory synaptic responses in the hippocampal CA1 area in vitro. Brain Res 500:333–342
102. Szatkowski M, Attwell D (1994) Triggering and execution of neuronal death in brain ischaemia: two phases of glutamate release by different mechanisms. Trends Neurosci 17:359–365
103. Iino M, Ozawa S, Tsuzuki K (1990) Permeation of calcium through excitatory amino acid receptor channels in cultured rat hippocampal neurones. J Physiol (Camb) 424:151–165
104. Nakanishi S (1992) Molecular diversity of glutamate receptors and implications for brain function. Science 258:597–603
105. Simon RP, Swan JH, Griffiths T, Meldrum BS (1984) Blockade of *N*-methyl-D-aspartate receptors may protect against ischemic damage in the brain. Science 226:850–852
106. Evans MC, Swan JH, Meldrum BS (1987) In: Raichle M (ed) 15th Princeton conference on cerebrovascular disease. Raven, New York, pp 27–36
107. Meldrum B, Evans M, Swan J, Simon RP (1987) Protection against hypoxic/ischaemic brain damage with excitatory amino acid antagonists. Med Biol 65:153–157
108. Swan JH, Evans MC, Meldrum BS (1988) Long-term development of selective neuronal loss and the mechanism of protection by 2-amino-7-phosphonoheptanoate in a rat model of incomplete forebrain ischaemia. J Cereb Blood Flow Metab 8:64–78
109. Swan JH, Meldrum BS (1990) Protection by NMDA antagonists against selective cell loss following transient ischemia. J Cereb Blood Flow Metab 10:343–351
110. Church J, Zeman S (1991) Ketamine promotes hippocampal CA1 pyramidal neuron loss after a short-duration ischemic insult in rats. Neurosci Lett 123:65–68
111. Buchan A, Li H, Pulsinelli WA (1991) The *N*-methyl-D-aspartate antagonist, MK-801, fails to protect against neuronal damage caused by transient, severe forebrain ischemia in adult rats. J Neurosci 11:1049–1056
112. McCulloch J, Ozyurt E, Park CK, Nehls DG, Teasdale GM, Graham DI (1993) Glutamate receptor antagonists in experimental focal cerebral ischaemia. Acta Neurochir (Suppl) 57:73–79
113. Sheardown MJ, Nielsen EØ, Hansen AJ, Jacobsen P, Honoré T (1990) 2,3-Dihydroxy-6-nitro-7-sulfamoyl-benzo(F)quinoxaline: a neuroprotectant for cerebral ischemia. Science 247:571–574
114. Sheardown MJ, Suzdak PD, Nordhoim L (1993) AMPA, but not NMDA, receptor antagonism is neuroprotective in gerbil global ischaemia, even when delayed 24h. Eur J Pharmacol 236:347–353
115. Judge ME, Sheardown MJ, Jacobsen P, Honoré T (1991) Protection against post-ischemic behavioral pathology by the α-amino-3-hydroxy-5-methyl-4-

isoxazolepropionic acid (AMPA) antagonist 2,3-dihydroxy-6-nitro-7-sulfamyol-benzo(f)quinoxaline (NBQX) in the gerbil. Neurosci Lett 133:291–294
116. Mosinger JL, Price MT, Bai HY, Xiao H, Wozniak DF, Olney JW (1991) Blockade of both NMDA and non-NMDA receptors is required for optimal protection against ischemic neuronal degeneration in the in vivo adult mammalian retina. Exp Neurol 113:10–17
117. Yassin M, Scholfield CN (1994) NMDA antagonists increase recovery of evoked potentials from slices of rat olfactory cortex after anoxia. Br J Pharmacol 111:1221–1227
118. Koh J-Y, Goldberg MP, Hartley DM, Choi DW (1990) Non-NMDA receptor-mediated neurotoxicity in cortical culture. J Neurosci 10:693–705
119. Garthwaite G, Garthwaite J (1989) Quisqualate neurotoxicity: a delayed, CNQX-sensitive process triggered by a CNQX-insensitive mechanism in young rat hippocampal slices. Neurosci Lett 99:113–118
120. Forsythe ID, Clements JD (1990) Presynaptic glutamate receptors depress excitatory monosynaptic transmission between mouse hippocampal neurones. J Physiol (Camb) 429:1–16
121. Baskys A, Malenka RC (1991) Agonists at metabotropic glutamate receptors presynaptically inhibit EPSCs in neonatal rat hippocampus. J Physiol (Camb) 444:687–701
122. Okamoto N, Hori S, Akazawa C, Hayashi Y, Shigemoto R, Mizuno N, Nakanishi S (1994) Molecular characterization of a new metabotropic glutamate receptor mGluR7 coupled to inhibitory cyclic AMP signal transduction. J Biol Chem 269:1231–1236
123. Saugstad JA, Kinzie JM, Mulvihill ER, Segerson TP, Westbrook GL (1994) Cloning and expression of a new member of the L-2-amino-4-phosphonobutyric acid-sensitive class of metabotropic glutamate receptors. Mol Pharmacol 45:367–372
124. Koh J-Y, Palmer E, Cotman CW (1991) Activation of the metabotropic glutamate receptor attenuates *N*-methyl-D-aspartate neurotoxicity in cortical cultures. Proc Natl Acad Sci USA 88:9431–9435
125. Koh J-Y, Palmer P, Lin A, Cotman CW (1991) A metabotropic glutamate receptor agonist does not mediate neuronal degeneration in cortical culture. Brain Res 561:338–343
126. Chiamulera C, Albertini P, Valerio E, Reggiani A (1992) Activation of metabotropic receptors has a neuroprotective effect in a rodent model of focal ischaemia. Eur J Pharmacol 216:335–336
127. Opitz T, Reymann KG (1993) (1S,3R)-ACPD protects synaptic transmission from hypoxia in hippocampal slices. Neuropharmacology 32:103–104
128. Schoepp DD, Johnson BG (1988) Excitatory amino acid agonist-antagonist interactions at 2-amino-4-phosphonobutyric acid-sensitive quisqualate receptors coupled to phosphoinositide hydrolysis in slices of rat hippocampus. J Neurochem 50:1605–1613
129. Schoepp DD, Johnson BG, Smith ECR, Mcquaid LA (1990) Stereoselectivity and mode of inhibition of phosphoinositide-coupled excitatory amino acid receptors by 2-amino-3-phosphonopropionic acid. Mol Pharmacol 38:222–228
130. Opitz T, Reymann KG (1991) Blockade of metabotropic glutamate receptors protects rat CA1 neurons from hypoxic injury. NeuroReport 2:455–457
131. Irving AJ, Schofield JG, Watkins JC, Sunter DC, Collingridge GL (1990) 1S,3R-ACPD stimulates and L-AP3 blocks Ca^{2+} mobilization in rat cerebellar neurons. Eur J Pharmacol 186:363–365
132. Baskys A (1992) Metabotropic receptors and "slow" excitatory actions of glutamate agonists in the hippocampus. Trends Neurosci 15:92–96
133. Hajos F, Garthwaite G, Garthwaite J (1986) Reversible and irreversible neuronal damage caused by excitatory amino acid analogues in rat cerebellar slices. Neuroscience 18:417–436
134. Garthwaite G, Garthwaite J (1986) Amino acid neurotoxicity: intracellular sites of calcium accumulation associated with the onset of irreversible damage to rat cerebellar neurones in vitro. Neurosci Lett 71:53–58

135. Tanabe Y, Masu M, Ishii T, Shigemoto R, Nakanishi S (1992) A family of metabotropic glutamate receptors. Neuron 8:169–179
136. Houamed KM, Kuijper JL, Gilbert TL, Haldeman BA, O'Hara PJ, Mulvihill ER, Almers W, Hagen FS (1991) Cloning, expression, and gene structure of a G protein-coupled glutamate receptor from rat brain. Science 252:1318–1321
137. Aramori I, Nakanishi S (1992) Signal transduction and pharmacological characteristics of a metabotropic glutamate receptor, mGluR1, in transfected CHO cells. Neuron 8:757–765
138. Abe T, Sugihara H, Nawa H, Shigemoto R, Mizuno N, Nakanishi S (1992) Molecular characterization of a novel metabotropic glutamate receptor mGluR5 coupled to inositol phosphate/Ca^{2+} signal transduction. J Biol Chem 267:13361–13368
139. Sahara Y, Westbrook GL (1993) Modulation of calcium currents by a metabotropic glutamate receptor involves fast and slow kinetic components in cultured hippocampal neurons. J Neurosci 13:3041–3050
140. Lester RAJ, Jahr CE (1990) Quisqualate receptor-mediated depression of calcium currents in hippocampal neurons. Neuron 4:741–749
141. Dolphin AC (1990) G Protein modulation of calcium currents in neurons. Annu Rev Physiol 52:243–255
142. Gray R, Johnston D (1987) Noradrenaline and β-adrenoceptor agonists increase activity of voltage-dependent calcium channels in hippocampal neurons. Nature 327:620–622
143. Lipscombe D, Bley K, Tsien RW (1988) Modulation of neuronal Ca channels by cAMP and phorbol esters. Soc Neurosci Abstr 14:153

Changes in Intracellular Ca^{2+} and pH of Hippocampal Slices in Response to Ischemia In Vitro

Naoshi Fujiwara, Takashi Abe, Yoshiko Ebine, and Koki Shimoji

Summary. Changes in intracellular Ca^{2+} ($[Ca^{2+}]_i$) and intracellular pH (pH_i) of hippocampal slice preparations in response to hypoxia/ischemia in vitro are measured by microscopic fluorometry using the fluorescent indicators 1-(2–5′-carboxyoxazol-2′-yl)-6-aminobenzofuran-5-oxy)-2-(2′-amino-5′-methylphenoxy)-ethane-*N*,*N*,*N*′,*N*′-tetraacetic acid (fura-2) and 2′,7′-bis(carboxyethyl)carboxyfluorescein (BCECF), respectively. $[Ca^{2+}]_i$ of the CA1 pyramidal cell layer showed a slow and small increase, and pH_i gradually declined, during hypoxia in vitro (in the presence of glucose). Both $[Ca^{2+}]_i$ and pH_i recovered by reoxygenation after hypoxia in vitro. In response to ischemia in vitro (oxygen/glucose deprivation), the slow and small increase in $[Ca^{2+}]_i$ was followed by a characteristic rapid increase, and the initial decrease in pH_i was followed by a gradual increase. $[Ca^{2+}]_i$ partially recovered, and pH_i substantially declined and then slowly recovered, but synaptic activity did not recover after reoxygenation and glucose resupply. These results suggested that the rapid increase in $[Ca^{2+}]_i$ is associated with ischemia-induced neuronal injury. On the other hand, effects of pH on changes in $[Ca^{2+}]_i$ induced by ischemia in vitro were investigated using the same slice model. The characteristic increase in $[Ca^{2+}]_i$ induced by ischemia in vitro was inhibited under the low pH condition (pH 6.8), which mimicked ischemic brain, while severe acidosis is thought to produce cell injury. Therefore, acidosis, if it is not severe enough to directly injure neurons and glial cells, may preserve hippocampal neurons from acute ischemic damage.

Key words. Intracellular Ca^{2+}—Intracellular pH—Hippocampal slice—Fura-2—BCECF—Hypoxia/ischemia—Acidosis

Introduction

In the ischemic brain, failure of energy metabolism and disturbance of cellular ionic and acid–base homeostasis are produced, which are thought to be the causes of acute neuronal injury [1]. Especially, elevation of intracellular free Ca^{2+} and tissue acidosis are proposed to be crucial in producing ischemic cell injury [2–5], but the detailed

Department of Anesthesiology, Niigata University School of Medicine, 1-757 Asahimachi-dori, Niigata, 951 Japan analysis of roles of intracellular Ca^{2+} concentration ($[Ca^{2+}]_i$) and acidosis for ischemia-induced neuronal damage is limited in in vivo study.

The hypoxia/ischemia model using brain slice preparations has been widely used in investigating the direct effects of hypoxia/ischemia on neuronal and glial activities [6,7], because extracellular conditions can be regulated by changing perfusion media as required. In addition, fluorophotometry using a Ca^{2+} indicator (1-(2-5′-carboxyoxazol-2′-yl)-6-aminobenzofuran-5-oxy)-2-(2′-amino-5′-methylphenoxy) ethane-*N,N,N′,N′*-tetraacetic acid, fura-2) or a pH indicator (2′,7′-bis(carboxyl)carboxyfluorescein, BCECF) enables us to measure intracellular Ca^{2+} $[Ca^{2+}]_i$) and intracellular pH (pH_i) of brain slices, respectively [8,9]. The in vitro model using brain slices is useful to analyze the mechanism of ischemic cell injury.

In this chapter, we demonstrate changes in $[Ca^{2+}]_i$ and pH_i of hippocampal slices in response to hypoxia/ischemia in vitro and discuss the roles of $[Ca^{2+}]_i$ and pH_i in ischemic neuronal damage. Effects of acid–base balance on $[Ca^{2+}]_i$ changes and cell viability during brain ischemia are also discussed.

Measurements of $[Ca^{2+}]_i$ and pH_i in Slice Preparations

Fluorophotometry of Slice Preparations Using Fluorescent Indicators

Fura-2 and BCECF have been most commonly used to measure $[Ca^{2+}]_i$ and pH_i, respectively. These fluorescent indicators are easily loaded into the intracellular space by incubating slice preparations with their acetoxymethyl esters in artificial cerebrospinal fluid (CSF) [8,9]. A hippocampal slice loaded with fura-2 or BCECF is superfused with artificial CSF in a measurement chamber placed on the stage of an inverted fluorescent microscope. The bottom of the recording chamber consists of a coverslip to transmit both excitation and emission light for microscopic fluorometry. A diagram of the experimental system is shown in Fig. 1.

For $[Ca^{2+}]_i$ measurements, a slice preparation was alternately irradiated with ultraviolet at 340 and 380 nm. Fluorescence emitted from slice preparations was passed through a bandpath filter (490–510 nm) and detected with a photomultiplier. Ratios of fluorescence intensities corresponding to excitation at 340 and 380 nm ($R_{340/380}$) were continuously recorded. The same fluorometry technique was used for pH_i measurements. Excitation wavelengths, 490 and 455 nm, and a bandpath filter (520–560 nm) were selected. Ratios of fluorescence intensities ($R_{490/455}$) were also continuously recorded.

Ischemia Model Using Slice Preparations (Ischemia In Vitro)

Perfusion systems using slice preparations enable us to control extracellular environments. Removal of oxygen or glucose from perfusion solution has been widely used as a hypoxia/ischemia model. In our experiments, oxygen deprivation (hypoxia in vitro) and oxygen/glucose deprivation (ischemia in vitro) were applied to slice preparations by superfusion with deoxygenated solutions that had been gassed with 95% N_2–5% CO_2. The temperature of the superfusate was controlled at 36–37°C. In this model, the tissue Po_2 of the slice preparation may be steeply decreased with depth, and the inside of the preparation is thought to be almost completely anoxic, even though deoxygenation of the bathing solution is not complete [10,11].

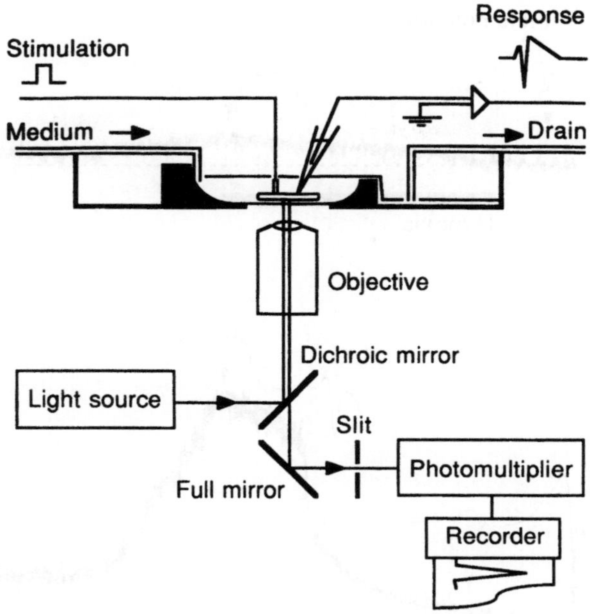

FIG. 1. A scheme of experimental system using brain slice preparation. A hippocampal slice was superfused with artificial cerebrospinal fluid (CFS) in the measurement chamber, the bottom of which consists of a coverslip. Fluorescence from the preparation was passed through an objective and a dichroic mirror and was detected with a photo multiplier. Measurement area in the view was selected using a slit. Fluorescence intensities for each excitation wavelength and their ratio are recorded with a three-channel recorder. Electrical activity can be also recorded by a ordinary method. As a standard artificial CSF, Krebs solution, consisting of 117 m*M* NaCl, 3.6 m*M* KCl, 1.2 m*M* NaH_2PO_4, 1.2 m*M* $MgCl_2$, 2.5 m*M* $CaCl_2$, 25 m*M* $NaHCO_3$, and 11 m*M* glucose and gassed with 95% O_2–5%CO_2, was used

In this ischemia model using slice preparations, electrical activities can be recorded at the same area where $[Ca^{2+}]_i$ is measured. In some experiments, $[Ca^{2+}]_i$ and extracellular field potentials were simultaneously recorded from the CA1 pyramidal cell layer of the hippocampal slices.

Changes in $[Ca^{2+}]_i$ in Response to Hypoxia/Ischemia In Vitro

As the elevation in $[Ca^{2+}]_i$ is thought to be crucial in producing ischemic neuronal damage [2,12,13], measurement of $[Ca^{2+}]_i$ in the brain tissue may be important for evaluating the cellular damage induced by ischemia. Microscopic fluorometry using fluorescent calcium indicators such as fura-2 enables us to measure changes in $[Ca^{2+}]_i$ in brain tissue [8].

Changes in $[Ca^{2+}]_i$ in the hippocampal CA1 region in response to hypoxia/ischemia in vitro are shown in Fig. 2. $[Ca^{2+}]_i$ slowly increased during hypoxia in vitro and decreased by reoxygenation. Field potentials evoked by stimulation of the Schaffer collaterals also recovered after reoxygenation. When slice preparations were loaded to

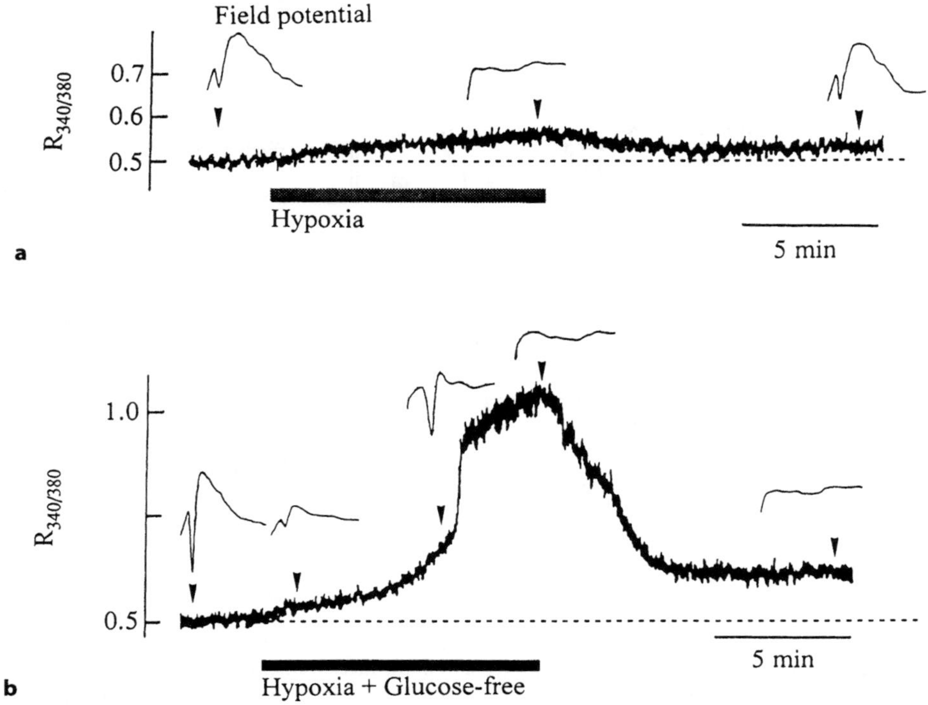

FIG. 2a,b. Changes in $[Ca^{2+}]_i$ and field potentials of CA1 pyramidal cell layer of hippocampal slices in response to hypoxia/ischemia in vitro. Field potentials were evoked by stimulation of the Schaffer collaterals and recorded at the time indicated by wedges below traces of potential. **a** Hypoxia in vitro (in the presence of 11 m*M* glucose). **b** Ischemia in vitro (hypoxia + glucose-free)

ischemia in vitro, $[Ca^{2+}]_i$ also slowly increased initially for several minutes such as was observed during hypoxia in vitro, but showed a characteristic rapid increase at 6–7 min. Field potentials transiently appeared coincidentally with the characteristic increase in $[Ca^{2+}]_i$. The $[Ca^{2+}]_i$ increased by ischemia in vitro was gradually reduced by resupply of oxygen and glucose, but not fully recovered. Field potentials did not recover after ischemia in vitro.

Occurrence of a hyperpolarization (5–15 mV) following a transient depolarization was reported in CA1 neurons in response to hypoxia, and these changes in membrane potential fully recovered after 10 min of hypoxia in vitro [13]. On the other hand, the characteristic increase in $[Ca^{2+}]_i$ coincided with a rapid membrane depolarization of CA1 pyramidal neurons in which membrane potential did not recover after oxygen and glucose were supplied again [14]. The increased $[Ca^{2+}]_i$ accompanied by a rapid depolarization may be crucial in producing ischemic neuronal damage, as Ca^{2+} accumulation is thought to cause neuronal injury by ischemia in vivo [2,3,12]. In the presence of glucose, cellular ionic balance may be maintained by anaerobic glycolysis, which may prevent a profound increase in $[Ca^{2+}]_i$ during 10 min of hypoxia in vitro.

Changes in pH_i in Response to Hypoxia/Ischemia In Vitro

Hypoxia and ischemia induce anaerobic glycolysis, which is thought to cause lactic acidosis in brain tissue. In most experiments using brain slices, however, the pH of solution used for hypoxia/ischemia applications is the same as that of oxygenated solutions. How does the pH_i of slice preparations change under hypoxic/ischemic conditions in uniform pH solutions?

Typical traces of pH_i of hippocampal slices exposed to hypoxia/ischemia in vitro are demonstrated in Fig. 3. With exposure to hypoxia in vitro, the pH_i of the CA1 pyramidal cell layer decreased but the rate of decline was suppressed in the latter period of hypoxia (Fig. 3a). After 10 min of hypoxia in vitro, pH_i rapidly recovered by reoxygenation.

Complicated changes in pH_i were observed in the slices exposed to ischemia in vitro (Fig. 3b). pH_i decreased during the initial period (until 4 min), but then increased and became close to or higher than the prehypoxic level at 10 min. After resupply of oxygen and glucose, pH_i profoundly decreased; this intracellular acidosis persisted several minutes, and pH_i then gradually recovered.

Lactate content in mouse hippocampal slices increased during hypoxia in vitro, and recovered to the prehypoxic level after reoxygenation [9]. The changes in pH_i are consistent with changes in lactate content in slice preparations loaded with hypoxia in vitro. The initial decrease and sequential increase in pH_i during ischemia in vitro may be also consistent with lactic contents in slice preparations. The production of lactate by glycolysis may also be suppressed in the glucose-deprived solution. Washing out with perfusate may prevent further accumulation of lactate in slice preparations during a continuing ischemic period. This may produce successive increases in pH_i. Indeed, lactate contents increased at 3 min, but decreased at 10 min (Fig. 4).

As intracellular acidosis following ischemia in vitro could not be explained only by the lactate content in slice preparations, other mechanisms should be considered as follows. The first, pH_i regulatory mechanisms, e.g., Na^+–H^+ exchange and HCO_3^--dependent acid extrusion, have been demonstrated in nervous systems [15–18]. As cellular ionic balance may be disturbed by ischemia in vitro, ion-exchange regulation of pH_i may produce intracellular acidification during recovery of the ionic balance. The second, excessive amount of glutamate release has been demonstrated in ischemic brain [19]. As glutamate uptake by active transport causes cellular acidification [20,21], cells in reoxygenated brain slices may be acidified by glutamate uptake, if glutamate is accumulated in the extracellular space during ischemia in vitro. The third, glucose metabolism, is thought to be recovered by reoxygenation and glucose resupply. Because a lot of energy is needed to recover the ion balance and for glutamate uptake, acid production (including CO_2) by glucose metabolism may be augmented in reoxygenated slices.

The recovery of pH_i may not necessarily reflect the recovery of neuronal functions. The membrane potential of CA1 neurons and synaptic activity in hippocampal slices recovered poorly after 10 min of ischemia in vitro [2,22–24]. Change in pH_i by itself may not cause neuronal damage by ischemia in vitro, but its changes reveal the cellular acid–base dynamism in slice preparations.

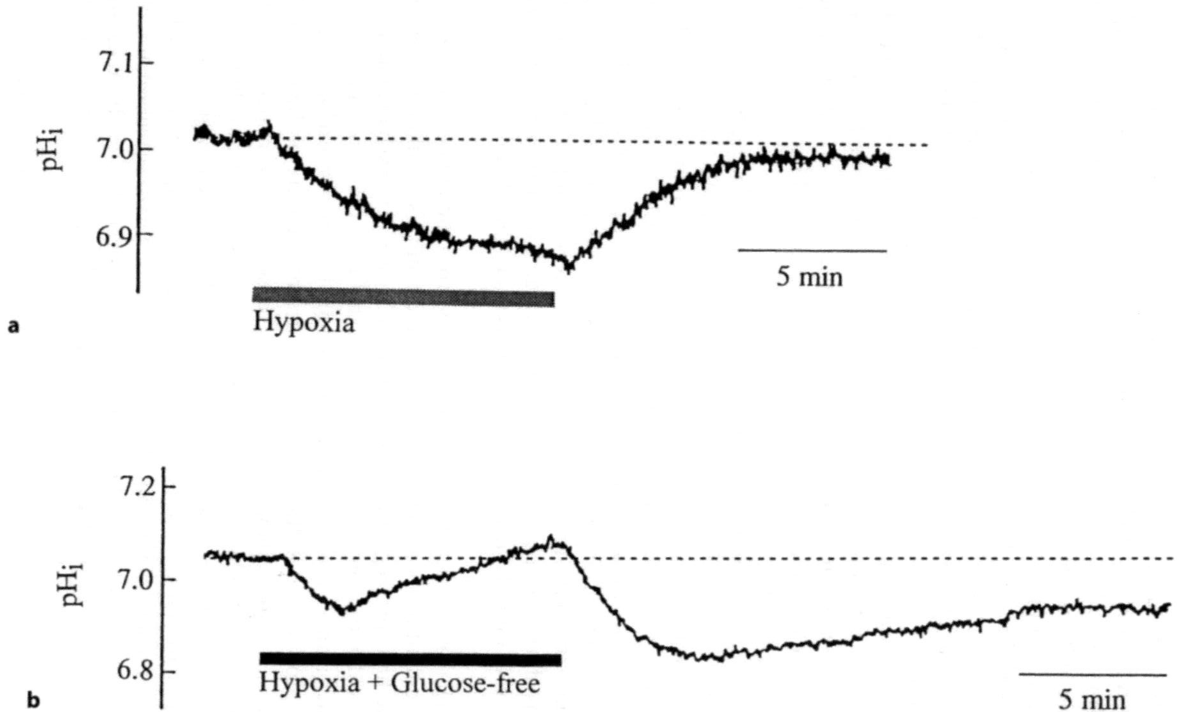

FIG. 3a,b. Typical traces in pH_i in CA1 pyramidal cell layer of hippocampal slices loaded to hypoxia/ischemia in vitro. **a** Hypoxia in vitro (in the presence of 11 m*M* glucose). **b** Ischemia in vitro (hypoxia + glucose-free)

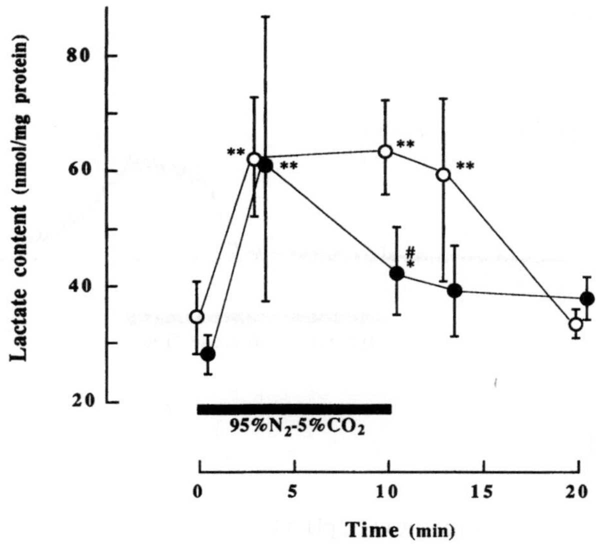

FIG. 4. Changes in lactate contents in mouse hippocampal slices loaded with hypoxia/ischemia in vitro. *Open circles*, 11 m*M* glucose; *solid circles*, glucose-free. Changes in pH_i in mouse hippocampal slices in response to hypoxia/ischemia in vitro were quite similar to those in rat hippocampal slices. Note that lactate contents did not increase after ischemia in vitro, which pH_i profoundly decreased in this period. *, $P < .05$, and **, $P < .01$, vs. the start (0 min) of hypoxia/ischemia in vitro; #, $P < .05$ vs. hypoxia in vitro at 10 min. This graph was produced using our numerical data in [9]

Effects of pH on Changes in $[Ca^{2+}]_i$ by Ischemia In Vitro

Because tissue acidosis is brought about by ischemia, the effects of acidosis on ischemic cell damage should be examined. To elucidate the effects of acidosis on neuronal damage induced by ischemia, changes in $[Ca^{2+}]_i$ in response to ischemia in vitro were observed in normal pH (pH 7.4), low pH (pH 6.8), and high pH (pH 7.7) solutions.

Typical changes in $[Ca^{2+}]_i$ in response to ischemia in vitro at various pHs are shown in Fig. 5. An initial slow increase in $[Ca^{2+}]_i$ followed by a characteristic rapid increase was observed in normal and high pH solutions. The characteristic rapid increase in $[Ca^{2+}]_i$ was suppressed in the low pH solution. In Ca^{2+}-free solution containing 0.5 m*M* *O,O'*-bis(2-aminoethyl)ethyleneglycol-*N,N,N',N'*-tetraacetic acid (EGTA), the characteristic rapid increase in $[Ca^{2+}]_i$ was also observed in solutions at normal and high pH, but not in the low pH solution. Under the oxygenated condition, no significant changes in $[Ca^{2+}]_i$ were induced by alteration of the pH of the perfusing solution.

Relative changes in $R_{340/380}$ at 10 and 15 min during ischemia in vitro are shown in Fig. 6. $R_{340/380}$ at 10 min was significantly ($P < .05$) lower in the low pH solution than in normal and high pH solutions regardless of whether the solution contained Ca^{2+}. At 15 min, the difference between $R_{340/380}$ increases in normal pH and low-pH solutions was insignificant. On the other hand, the increase in $R_{340/380}$ at 15 min in the low pH solution was significantly less than those in normal and high pH solutions under the Ca^{2+}-free condition (0 m*M* Ca^{2+} and 0.5 m*M* EGTA). In the high pH solution, increase

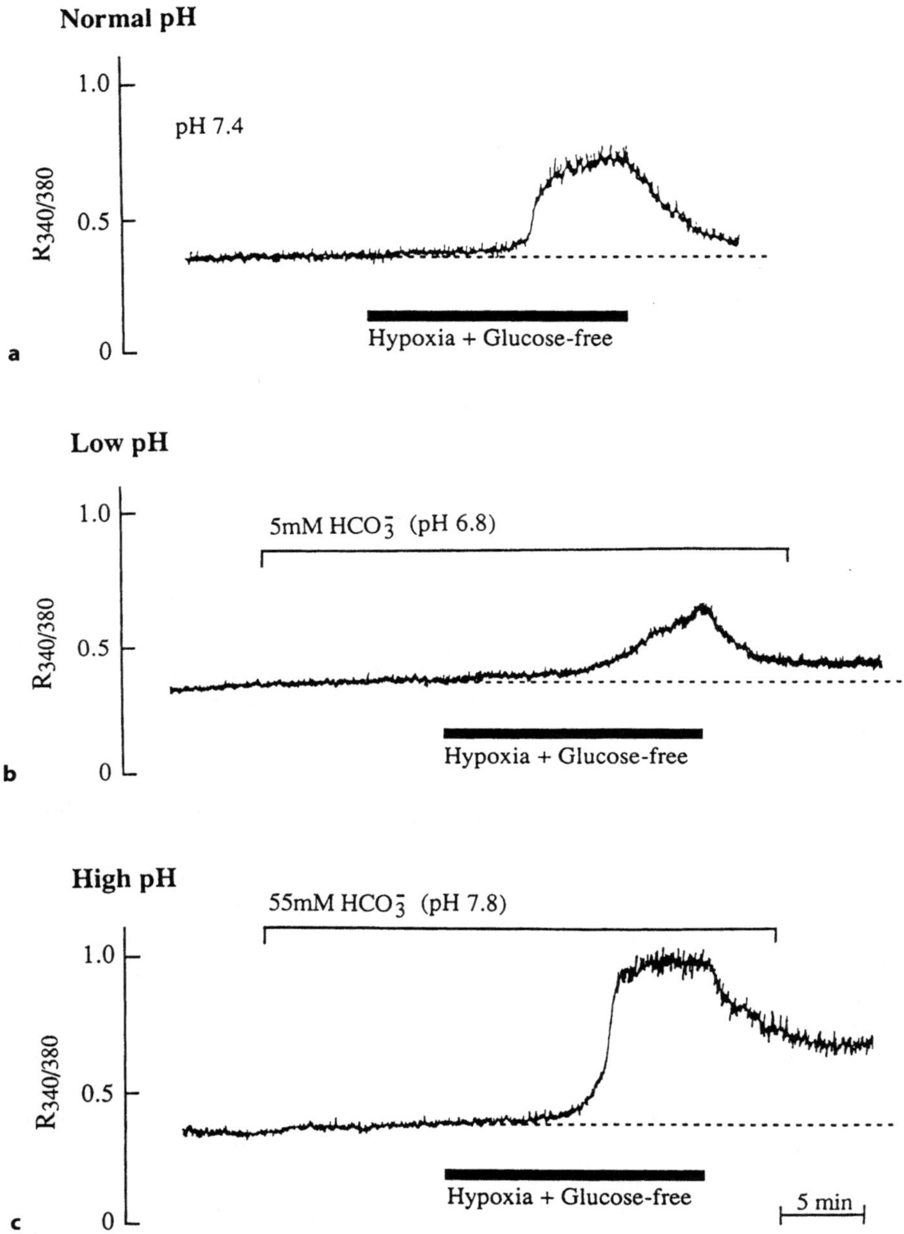

Fig. 5a–c. Typical traces of changes in $R_{340/380}$ in slice preparations loaded with ischemia in vitro with normal (**a**), low (**b**), and high pH (**c**) solutions. Normal Krebs solution was used for the normal pH solution with $NaHCO_3$ concentration at 25 m*M*; $NaHCO_3$ concentration was 5 m*M* in the low pH solution and 55 m*M* in the high pH solution. Tonicity was adjusted with NaCl

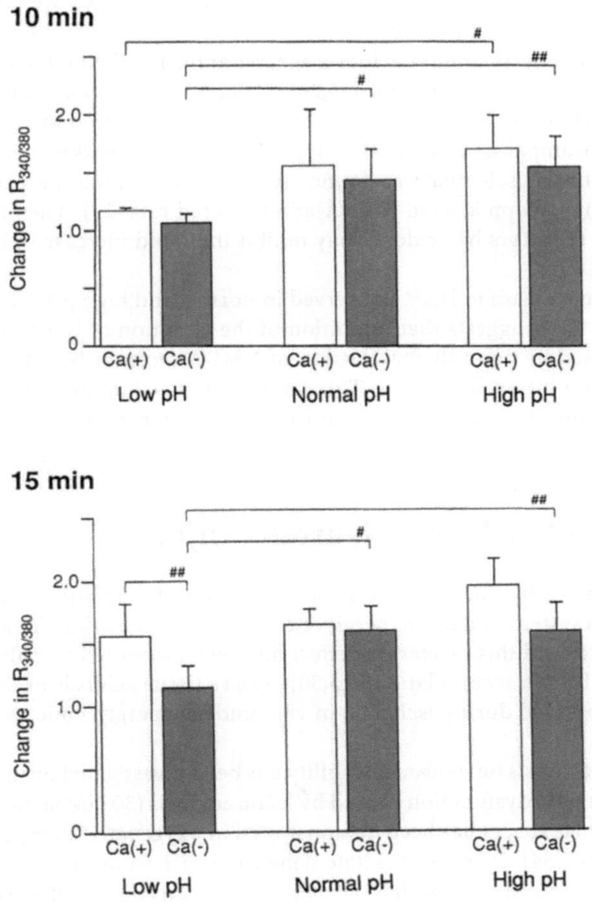

FIG. 6. Changes in $[Ca^{2+}]_i$ of CA1 pyramidal cell layer in response to ischemia in vitro in normal (2.5 m*M*) Ca^{2+} and Ca^{2+}-free solutions at various pHs. *Longitudinal axis* shows the ratio of $R_{340/380}$ at 10 min (*top*) and 15 min (*bottom*) to basal $R_{340/380}$ at the start of ischemia in vitro in 2.5 m*M* Ca^{2+} (Ca(+), white bars) and Ca^{2+}-free (0 m*M* Ca^{2+} with 0.5 m*M* ethylenegly-coltetraacetre acid (EGTA), Ca(−), *stippled bars*) solutions. Normal pH, 7.42 ± 0.07; low pH, 6.81 ± 0.08; high pH, 7.74 ± 0.06; #, $P < .05$; ##, $P < .01$. These graphs were produced using our numerical data in [39]

in $R_{340/380}$ at 15 min under the Ca^{2+}-free condition without Ca^{2+} was significantly less than that in the 2.5 m*M* Ca^{2+} solution.

The results indicate that the rise in $[Ca^{2+}]_i$ in response to ischemia in vitro is inhibited in a low pH solution and is accelerated in a high pH solution as compared to that in a solution at normal pH. As the characteristic rapid increase in $[Ca^{2+}]_i$ may be crucial in producing ischemic neuronal damage, neurons in hippocampal slices in the low pH solution may be more resistant to ischemia in vitro than those in normal and high pH solutions. The elevation of $[Ca^{2+}]_i$ observed at normal and high pH in the absence of Ca^{2+} indicates that intracellular calcium is an important source of the increase in $[Ca^{2+}]_i$ in response to ischemia in vitro. Therefore, both Ca^{2+} influx from the extracellular space and Ca^{2+} release from intracellular sites are partially inactivated under the low pH condition.

The protective effects of mild acidosis against anoxia and glutamate toxicity were demonstrated in cultured neurons [25,26]. Calcium conductance mediated by *N*-methyl-D-aspartate (NMDA) receptors was found to be reduced in cultured cortical [27] and hippocampal neurons [28] in low pH solutions. Marked delay in calcium influx during brain ischemia was found in rats with acidosis induced by hyperglycemia and hypercapnia, as in NMDA-administered rats [29]. Therefore, inactivation of NMDA receptors by acidosis may inhibit the rapid increase in $[Ca^{2+}]_i$ caused by ischemia in vitro.

However, the increase in $[Ca^{2+}]_i$ observed in normal and high pH solutions even in the absence of Ca^{2+} suggests that inhibition of the elevation of $[Ca^{2+}]_i$ in the low pH solution is caused by not only inactivation of NMDA receptor but also inhibition of Ca^{2+} release from intracellular sites. This acidosis-induced inhibition of intracellular Ca^{2+} mobilization may play an important role in resistance of neurons to ischemic damage.

Acidosis and Ischemic Neuronal Injury

Acidosis is commonly believed to be one of the causes of ischemic cell damage in the central nervous system. Indeed, hyperglycemia augments the brain damage caused by ischemia in vivo, and this deleterious effect has been attributed to cellular acidification caused by lactate accumulation [4,5,30]. Severe tissue acidosis near or below pH 6.0 has been observed during ischemia in vivo under hyperglycemic conditions [31–33].

The effect of acidosis on neuronal viability has been investigated by in vitro studies. Irreversible synaptic dysfunction caused by lactic acidosis (30 m*M* lactic acid, pH 5.99 in the extracellular space) has been demonstrated in oxygenated slice preparations of rat hippocampus [34]. Release of lactate dehydrogenase from hippocampal cell cultures, which reflects cell damage, has been observed in on HCl-acidified solution at pH 6.0, even in the presence of both oxygen and glucose [35]. In addition, it has been shown that volume changes and the viability of cultured glial cells are significantly deteriorated at pH 6.2 or lower, but not at pH 6.8 [36]. These findings suggest that severe acidosis may directly induce cell injury, but mild acidosis by itself may not produce cell destruction.

However, acidosis in ischemic brain is thought to be not so severe under normoglycemic conditions [32,33]. How does such mild acidosis affect ischemia-induced changes of neuronal activities? It has been observed that recovery of synaptic function following hypoxia in vitro is improved in low pH (pH 6.8–6.9) solutions prepared with lactic acid [37]. The ischemia-induced increase in $[Ca^{2+}]_i$ was inhibited in the low pH (pH 6.8) solution, as mentioned previously [38]. Reduction of neuronal excitability by mild acidosis (pH 7.1) has been demonstrated in hippocampal neurons [39]. This acidosis-induced inhibition of neuronal excitability may slow down the energy depletion caused by ischemia. Moreover, protection of cultured neurons from anoxic damage and glutamate excitotoxicity has been demonstrated under acidic conditions [26,27,40]. It is reported that acidosis reduced infarct volume in ischemic rat brains with maximal protection at brain pH 6.8, but the effect was lost at pH 6.5 [41]. From these findings, one may propose the following hypothesis: mild acidosis may prevent neuronal injury induced by ischemia or glutamate excitotoxicity.

Conclusion

In normal pH solution (pH 7.4), $[Ca^{2+}]_i$ of the CA1 pyramidal cell layer showed a slow and small increase, and pH_i gradually declined in response to hypoxia in vitro. $[Ca^{2+}]_i$, pH_i, and synaptic activity recovered after hypoxia in vitro. In response to ischemia in vitro, the slow and small increase in $[Ca^{2+}]_i$ was followed by a characteristic rapid elevation. The initial decrease in pH_i was followed by a gradual increase during ischemia in vitro. pH_i substantially declined and then slowly recovered after reoxygenation and glucose resupply. Synaptic activity recovered after hypoxia in vitro, but did not recover after ischemia in vitro.

On the other hand, the characteristic increase in $[Ca^{2+}]_i$ induced by ischemia in vitro was inhibited under the low pH condition (pH 6.8) that mimicked an ischemic brain, while severe acidosis is thought to cause cell injury. Therefore, acidosis, if not severe enough to directly injure neurons and glial cells, may preserve hippocampal neurons from acute ischemic damage.

Acknowledgment. This work was supported in part by a grant-in-aid (07407043) from the Japanese Ministry of Education, Science and Culture.

References

1. Siesjö BK (1981) Cell damage in the brain: a speculative synthesis. J Cereb Blood Flow Metabol 1:155–185
2. Deshpande JK, Siesjö BK, Wieloch T (1987) Calcium accumulation and neuronal damage in the rat hippocampus following cerebral ischemia. J Cereb Blood Flow Metab 7:89–95
3. Siesjö BK, Bengtsson F (1989) Calcium fluxes, calcium antagonists, and calcium-related pathology in brain ischemia, hypoglycemia, and spreading depression: a unifying hypothesis. J Cereb Blood Flow Metab 9:127–140
4. Rehncrona S, Rosén I, Siesjö BK (1981) Brain lactic acidosis and ischemic cell damage. 1. Biochemistry and neurophysiology. J Cereb Blood Flow Metab 1:297–311
5. Kalimo H, Rehncrona S, Söderfeldt B, Olsson Y, Siesjö BK (1981) Brain lactic acidosis and ischemic cell damage. 2. Histopathology. J Cereb Blood Flow Metab 1:313–327
6. Lye RH, Clough PH (1989) The applicability of brain slice methodology to the study of focal cerebral ischemia in the rat. J Neurosci Methods 28:1–6
7. Shimoji K, Fujiwara N, Endoh H, Abe T (1991) Experimental ischemia model using brain slices (in Japanese). Byotai Seiri Pathophysiology 10:343–350
8. Kudo Y, Ito K, Miyakawa H, Izumi Y, Ogura A, Kato H (1987) Cytoplasmic calcium elevation in hippocampal granule cell induced by perforant path stimulation and L-glutamate application. Brain Res 407:168–172
9. Fujiwara N, Abe T, Endoh H, Warashina A, Shimoji K (1992) Changes in intracellular pH of mouse hippocampal slices responding to hypoxia and/or glucose depletion. Brain Res 572:335–339
10. Fujii T, Baumgärt H, Lübber DW (1982) Limiting section of guinea pig olfactory cortical slices studies from tissue PO_2 values and electrical activities. Pflügers Arch 393:83–87
11. Bingmann D, Kolde G (1982) PO_2-profiles in hippocampal slice of the guinea pig. Exp Brain Res 48:89–96
12. Choi DW (1988) Calcium-mediated neurotoxicity: relationship to specific channel type and role in ischemic damage. Trends Neurosci 11:465–469
13. Fujiwara N, Higashi H, Shimoji K (1987) Effect of hypoxia on rat hippocampal neurones in vitro. J Physiol 348:131–151

14. Higashi H, Kudo Y, Nishi S (1990) Dysfunction of hippocampal pyramidal cells induced by deprivation of oxygen and glucose. Neurosci Res Suppl 11:S18
15. Chesler M (1990) The regulation and modulation of pH in the nervous system. Prog Neurobiol 34:401–427
16. Chesler M (1987) pH regulation in the vertebrate central nervous system: microelectrode studies in the brain stem of the lamprey. Can J Physiol Pharmacol 65:986–993
17. Schlue WR, Deitmer JW (1987) Direct measurement of intracellular pH in identified glial cells and neurons of the leech central nervous system. Can J Physiol Pharmacol 65:978–985
18. Schlue WR, Thomas RC (1985) A dual mechanism for intracellular pH regulation by leech neuron. J Physiol 364:327–338
19. Benveniste H, Drejer H, Schousboe A, Diemer NH (1984) Elevation of the extracellular concentrations of glutamate and aspartate in rat hippocampus during transient cerebral ischemia monitored by intracerebral microdialysis. J Neurochem 43:1369–1374
20. Amato A, Ballerini L, Attwell D (1994) Intracellular pH changes produced by glutamate uptake in rat hippocampal slices. J Neurophysiol (Bethesda) 72:1686–1695
21. Bouvier M, Szatkowski M, Amato A, Attwell D (1992) The glial cell glutamate uptake carrier countertransports pH-changing anions. Nature 360:471–474
22. Dong WQ, Schurr A, Reid K, Shield CB, West CA (1988) The rat hippcampal slice preparation as an in vitro model of ischemia. Stroke 19:498–502
23. Schurr A, West CA, Reid KH, Tseng MT, Reiss ST, Rigor BM (1987) Increased glucose improves recovery of neuronal function after cerebral hypoxia in vitro. Brain Res 421:135–139
24. Schurr A, West CA, Rigor BM (1989) Electrophysiology of energy metabolism and neuronal function in the hippocampal slice preparation. J Neurosci Methods 28:7–13
25. Kaku DA, Giffard RG, Choi DW (1993) Neuroprotective effects of glutamate antagonists and extracellular acidity. Science 260:1516–1518
26. Tombaugh GC, Sapolsky RM (1990) Mild acidosis protects hippocampal neurons from injury induced by oxygen and glucose deprivation. Brain Res 506:343–345
27. Giffard PG, Monyer H, Christine CW, Choi DW (1990) Acidosis reduces NMDA receptor activation, glutamate neurotoxicity, and oxygen-glucose deprivation neuronal injury in cortical cultures. Brain Res 506:339–342
28. Tang CM, Dichter M, Morad M (1990) Modulation of the *N*-methyl-D-aspartate channel by extracellular H^+. Proc Natl Acad Sci USA 87:6445–6449
29. Kristián T, Katsura K, Gidö G, Siesjö BK (1994) The influence of pH on cellular calcium influx during ischemia. Brain Res 641:295–302
30. Gardiner M, Smith ML, Kågström E, Shohami E, Siesjö BK (1982) Influence of blood glucose concentration on brain lactate accumulation during severe hypoxia and subsequent recovery of brain energy metabolism. J Cereb Blood Flow Metabol 2:429–438
31. Combs DJ, Dempsey RJ, Maley M, Donaldson D, Smith C (1990) Relationship between plasma glucose, brain lactate, and intracellular pH during cerebral ischemia in gerbils. Stroke 21:936–942
32. Katsura K, Ekholm A, Asplund B, Siesjö BK (1991) Extracellular pH in the brain during ischemia: relationship to the severity of lactic acidosis. J Cereb Blood Flow Metab 11:597–599
33. Widmer H, Abiko H, Faden AI, James TL, Weinstein PR (1992) Effects of hyperglycemia on the time course of changes in energy metabolism and pH during global cerebral ischemia and reperfusion in rats: correlation of ^{1}H and ^{31}P NMR spectroscopy with fatty acid and excitatory amino acid levels. J Cereb Blood Flow Metab 12:456–468
34. Walz W, Harold DE (1990) Brain lactic acidosis and synaptic function. Can J Physiol Pharmacol 68:164–169
35. Tombaugh GC, Sapolsky RM (1990) Mechanistic distinctions between excitotoxic and acidotic hippocampal damage in an in vitro model of ischemia. J Cereb Blood Flow Metab 10:527–535

36. Staub F, Baethmann A, Peters J, Weigt H, Kempski O (1990) Effects of lactacidosis on glial cell volume and viability. J Cereb Blood Flow Metab 10: 866–876
37. Schurr A, Dong WQ, Reid KH, West CA, Rigor BM (1988) Lactic acidosis and recovery of neuronal function following cerebral hypoxia in vitro. Brain Res 438:311–314
38. Ebine Y, Fujiwara N, Shimoji K (1994) Mild acidosis inhibits the rise in intracellular Ca^{2+} concentration in response to oxygen-glucose in rat hippocampal slices. Neurosci Lett 168:155–158
39. Balestrino M, Somjen GG (1988) Concentration of carbon dioxide, intrastitial pH and synaptic transmission in hippocampal formation of the rat. J Physiol 396:247–266
40. Kaku DA, Giffard RG, Choi DW (1993) Neuroprotective effects of glutamate antagonists and extracellular acidity. Science 260:1516–1518
41. Simon PR, Niiro M, Gwinn R (1993) Brain acidosis induced by hypercarbic ventilation attenuates focal ischemia injury. J Pharmacol Exp Ther 267:1428–1431

Role of Nitric Oxide in Ischemic Brain Damage

TORU MATSUI and TAKAO ASANO

Summary. Our results on the role of nitric oxide (NO) in cellular mechanisms of ischemic brain damage are as follows. (1) Repeated i.p. administration of N^G-nitro-L-argininc (L-NNA) mitigated rat brain edema or infarction following permanent middle cerebral artery (MCA) occlusion at a dosage of 0.01–1 mg/kg. (2) In brain microvessels obtained from the affected hemisphere, Ca^{2+}-dependent constitutive nitric acid synthase (NOS) (E-c-NOS) was activated consistently at an early stage. On the other hand, Ca^{2+}-independent inducible NOS (i-NOS) was activated at 4 h and 24 h after MCA occlusion. (3) In the cerebral cortex, only Ca^{2+}-dependent constitutive NOS (N-c-NOS) was activated during the first 4 h of MCA occlusion in rats. (4) A2-h reperfusion following a 2-h MCA occlusion caused a significant increase in E-c-NOS and N-c-NOS activity without any apparent alterations in i-NOS activity. (5) NO concentration in the affected hemisphere was remarkably increased at 15–30 min and subsequently at 3–4 h after MCA occlusion. (6) Restoration of blood flow for 2 h after a 2-h MCA occlusion resulted in an enhanced NO concentration at 1–1.5 h after recirculation. (7) Administration of L-NNA (1 mg/kg i.p.) diminished the increments of NO concentration during ischemia and reflow, leading to its neuroprotection. Although it remains to be determined whether NO is neurotoxic or neuroprotective in cerebral ischemia, these results directed us to the conclusion that NOS activation in brain microvessels and cerebral cortex may play a role in the cellular mechanisms underlying ischemic brain injury.

Key words. Nitric oxide—Neuronal constitutive nitric oxide synthase—Endothelial constitutive nitric oxide synthase—Inducible nitric oxide synthase—Middle cerebral artery occlusion—Nitric oxide-sensitive electrode

Introduction

The simple gas nitric oxide (NO) was first identified as endothelium-derived relaxing factor (EDRF) in the vasculature. The accumulated evidence demonstrates that it is involved in many physiological conditions including central and peripheral

Department of Neurosurgery, Saitama Medical Center/School, 1981 Kamoda, Kawagoe, Saitama 350, Japan vasorelaxation, control of platelet function, cell proliferation, and neurotransmission [1,2]. This chapter focuses on the hypothesis that NO-induced neurotoxicity may be linked to ischemic neuronal cell damage [3], which has been deduced from the following experimental results:

1. Cerebral ischemia induces an increase in the extracellular concentration of glutamate and aspartate, which stimulate the glutamatergic receptors.
2. Excessive stimulation of the glutamatergic *N*-methyl-D-aspartate (NMDA) receptor leads to an remarkable production of NO via stimulation of NO synthase (NOS) in the brain tissue.
3. In vitro experiments demonstrate that NO is implicated in the glutamatergic neurotoxicity induced by NMDA.

Accordingly, many laboratories have examined the role of NO in the pathophysiology of ischemic cerebral injury in vivo. They administered many types of NOS inibitors to the models of cerebral ischemia. However, the efficacy of NOS inhibitors in focal cerebral ischemic models have generated widely disparate findings, from neuroprotection from deleterious outcome. The authors thus aimed at reviewing hitherto accumulated results, including our own, and at answering the question, "What is the truth about the role of NO in cerebral ischemia?"

This article consists of six parts: an introduction of the NO-producing system, the linkage of NO-induced cytotoxicity to ischemic neuronal damage, characterization of commercialized NOS inhibitors, review of the results on the efficacy of NOS inhibition in cerebral ischemia, review of direct and indirect methods of NO measure in the ischemic brain, and our conclusions.

Nitric Oxide Synthase

To date, at least three distinct clones of NOS enzymes have been obtained from the brain, endothelium, macrophage, smoth muscle cells, chondrocytes, hepatocytes, and other locations [4–22]. The deduced amino acid sequences of these NOS clones have predicted binding sites for reduced nicotinamide adenine dinucleotide (NADPH), flavin mononucleotide (FMN), flavin adenine dinucleotide (FAD), and (Calmodulin) (CaM). All the NOS clones have been demonstrated to required L-arginine, molecular O_2, and NADPH as cosubstrates and FMN, FAD, heme, and tetrahydrobiopterin as cofactors in the reaction [23,24]. The NOS family consists of the following three members: type I, in brain and neurons (N-c-NOS; M_r = 150–160 kDa) [4,25–28], and type III, in endothelial cells (E-c-NOS; M_r = 135 kDa) [29–31], are constitutively expressed and regulated by Ca^{2+} and CaM. This type contains the myristoylation domain for membrane binding. Type I c-NOS has been also found in skeletal muscle, pancreas, kidney, liver, stomach, and womb [23,32]. N- and E-c-NOS produce NO transiently in the agonist-induced increase in intracellular calcium. Type II is Ca^{2+}-independent i-NOS (M_r = 130 kDa), expressed by stimulation of endotoxins or cytokines in numerous cells such as macrophage, glia, and endothelium [33–38]. The i-NOS causes the sustained production of NO [23]. E-c-NOS exists predominantly in the particulate fraction (insoluble, membrane-bound). In contrast, neuronal c-NOS and i-NOS are found mainly in the soluble fraction [39].

NO-Induced Neurotoxicity and Its linkage to Cerebral Ischemia

Dawson et al. [40] first reported that a NOS inhibitor prevents neurotoxicity elicited by NMDA and related excitatory amino acids in primary cortical cultures. This effect was reversed by L-arginine, and depletion of the culture medium of arginine or use of arginine-free growth medium attenuated NMDA toxicity. In addition, hemoglobin, which complexes NO, prevented the neurotoxic effects of both NMDA and sodium nitroprusside (SNP), an NO donor [1,3,40,41]. Because cerebral ischemia causes an accumulation of extracellular glutamate [42], this leads to the increase in intracellular calcium concentration ($[Ca^{2+}]_i$) involved in neuronal cell damage. Based on the concept that NOS may be highly activated via an increase in $[Ca^{2+}]_i$, NO production over the physiological levels is thought to participate in ischemic neuronal damage. However, it should be added that a recent report [43] presented results inconsistent with those of Dawson et al.

Characteristics of NOS Inhibitors

To examine the foregoing hypothesis, many researchers administered commercially obtained NOS inhibitors to the experimental model of cerebral ischemia. In this section, these pharmacological agents are introduced. One of the substrates for NOS is L-arginine, and as NOS requires several cofactors in the reaction steps, certain N^G-monosubstituted L-arginine analogs and cofactor-antagonists or cofactor-synthesizing enzyme inhibitors have proven to possess an ability of variable potency to suppress NOS activity.

Among N^G-monosubstituted L-arginine analogs, N^G-nitro-L-arginine (L-NNA), N^G-nitro-L-arginine methyl ester (L-NAME), N^G-monomethyl-L-arginine (L-NMMA), N^G-amino-L-arginine (L-AA), and N^G-iminoethyl-L-ornithine (L-NIO) have been extensively characterized [44–48]. Although L-NNA [49] and N^G-cyclopropyl-L-arginine (L-CPA) [50] have been reported to selectively abolish N-c-NOS activity, the former inhibits E-c-NOS [51] and the latter [52] macrophage i-NOS. L-NNA inhibits N-c-NOS irreversibly [53] but E-c-NOS reversibly [44,51], and macrophage i-NOS [49,53], L-NIO, and L-NMMA reversibly inhibit E-c-NOS [44], but the i-NOS inhibition of L-NIO has been shown to be irreversible [45]. L-CPA is a reversible inhibitor and N^G-allyl-L-arginine (L-ALA) is both reversibly and irreversibly inhibiting to macrophage i-NOS [52]. Aminoguanidine (AG), *N*, *N*′-diaminoguanidine, methylguanidine, and 1,1-dimethylguanidine are known to reduce endothelial c-NOS or macrophage i-NOS activity [54]. In particular, AG has been found to preponderantly restrain i-NOS activity [54,55].

We investigated the concentration required for a 50% inhibition (IC_{50}) with well-known NOS inhibitors by means of monitoring the conversion of L-[^{3}H]arginine to L-[^{3}H]citrulline [56]. Rat brain cortical cytosol was used as the enzyme source of N-c-NOS. It was shown that the IC_{50} values of L-NNA, L-NAME, and L-NMMA in this fraction were entirely similar, respectively, to those in the brain cortical synaptosomes (unpublished data). It seems reasonable, therefore, to conceive that the observed NOS activities under our conditions originate mainly from N-c-NOS. A homogenate of cultured bovine pulmonary artery endothelial cells was used to assess E-c-NOS activity. To prepare the iNOS source, the rat lung cytosol fraction was collected at 6h after i.p. injection of lipopolysaccharide (10mg/kg) as reported previously [57]. It was shown that L-NNA is the most potent and AG is the weakest inhibitor of N-c-NOS in these four compounds. However, L-NNA had a relatively high affinity for E-c-NOS. 7-Nitro indazole (7-NI) and related indazoles have been reported to inhibit cerebellar c-NOS, exhibit antinociceptive activity [58–60], and relax smooth muscle [61].

Cloricromene, a coumarine derivative [62], pyrrolidine dithiocarbamate, an inhibitor of nuclear factor (NF) κβ [37], 3-amino-1,2,4-triazole, a catalase/peroxidase inhibitor [63], chlorpromazine [64], dihydropyridine antagonists, and agonists of calcium channels [65] have been reported to inhibit the induction of i-NOS. However, chlorpromazine also inhibits c-NOS activity [64]. These compounds appear to act as nonspecific inhibitors, and their potency has been shown to be relatively weak in comparison with the N^G-monosubstituted L-arginine derivatives. To our knowledge, no inhibitors specific to either E-c-NOS, *N*-c-NOS, or i-NOS have been developed, and research associated with this field is still anticipated.

Effects of NOS Inhibition Against Focal Cerebral Ischemia

Hitherto accumulated experimental evidence on the effects of NOS inhibitors against focal ischemia in rodent has shown disparate findings. Ameliorative effects have been seen in mice and rats when a low dose of NOS inhibitors had been repetitively administered i.p. during a 1-day period following ischemia [66–70]. In addition, NOS inhibitors have been shown to protect against hypoxic brain injury [71] and cold-induced brain edema [72] in rats. However, acute administration of high-dose NOS inhibitors pre- and/or postischemically have not elicited ameliorative effects in rats [73,74].

Yamamoto et al. [75] reported that infused L-NNA aggravates infarct volume after middle cerebral artery (MCA) occlusion in spontaneously hypertensive rats (SHR) [75]. They infused L-NNA for 1h (2.4mgkg−1h−1) immediately following occlusion of the MCA. This treatment caused significant increases in both mean arterial blood pressure (MABP) and cerebrovascular resistance without alterations in regional cerebral blood flow (rCBF). Here, one should ask why these contradictory results were obtained in the same model of cerebral ischemia.

Difference in the Dose of NOS Inhibitor Administration

We first did a dose-dependent study of L-NNA against brain edema formation at 72h post MCA occlusion in rats [76]. As shown in Fig. 1, the treatment of ischemic animals with very low doses of L-NNA (0.03mg/kg) at 5min and at 3, 6, and 24h after ischemia (total, 0.12mg/kg) exerted the most profound ameliorative effects. Statistically significant protective effects were observed in the 0.01–1mg/kg groups. In parallel, we also investigated the effects of L-NNA on mean systemic arterial pressure (MSAP) and CBF, which is measured with the hydrogen clearance method. When more than 3mg/kg of LNA was given i.p. to normal rats, MSAP increased by 20%–30% of the baseline value but CBF considerably decreased (Fig. 2). These findings suggested that a high dose of NOS inhibitor is likely to cause cerebral microcirculatory disturbance and may partly explain the results obtained from the in vivo dose-escalation study.

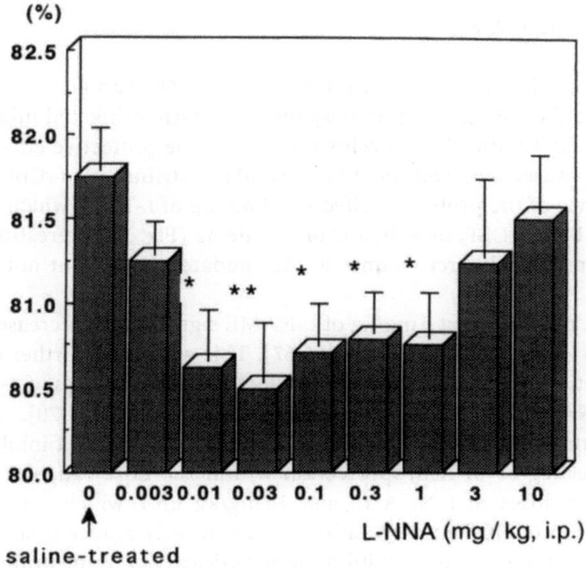

Fig. 1. Dose-dependent effects of N^G-nitro-L-arginine (L-NNA) on the affected hemispheric water contents after permanent left middle cerebral artery (MCA) occlusion in rats. Either saline (1 ml/kg) or L-NNA (0.003–10 mg/kg) was given i.p. to each animal four times at 5 min and at 3, 6, and 24 h after the onset of cerebral ischemia. The animals were killed 72 h after ischemia and the left hemispheric water contents measured by a dry-wet method. Each bar (*shaded*, water content) represents the mean ± SEM. *, $P < 0.05$ vs. saline-treated group; **, $P < 0.01$ vs. saline-treated group (ANOVA, followed by Dunnett test)

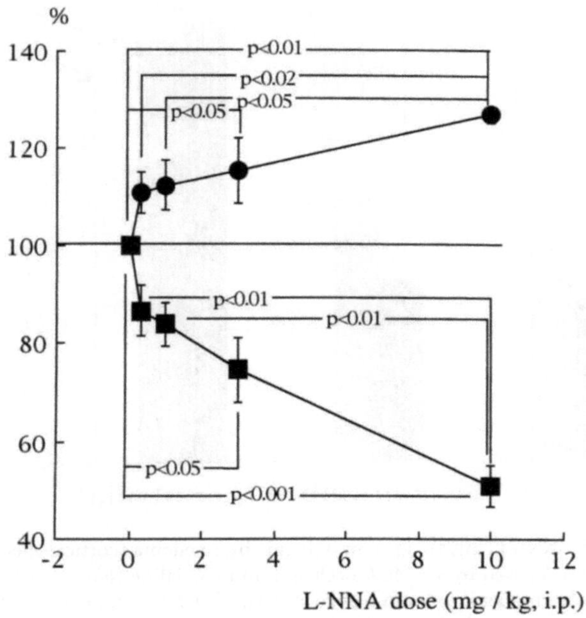

Fig. 2. Effects of L-NNA administration on mean arterial blood pressure (MABP, *circles*) and regional cerebral blood flow (rCBF, *squares*). These two factors showed the paradoxical feature. Each bar represents the mean ± SEM. *P* values were obtained from Wilcoxan rank test

Nishikawa et al. [77] have demonstrated that L-NAME (10 mg/kg i.v.), which completely abolishes the cortical NOS activity for at least 4h, mitigated injury within the caudate nucleus at 4h after MCA occlusion in cats. The protective effect of L-NAME is unlikely to be caused by elevation of MABP and redistribution of rCBF [77]. We also obtained evidence of the protective effects of 1 mg/kg of L-NNA, which is considered to not alter MABP or rCBF, in ischemic brain edema (Fig. 3). Interestingly, L-NAME (10 mg/kg i.v.) reduces infarct volume in the caudate nucleus but not in the cortex [77].

Buisson et al. reported that 3 mg/kg of L-NAME significantly decreased infarct size in the cortex as well as in the striatum [66,67]. This group has further examined the detailed correlation of the neuroprotective effects of various doses of L-NNA and the degree of L-NNA-induced inhibition of cortical NOS activity [70]. These reports suggested that there was a good correlation between the extent of inhibition of NOS activity and the degree of neuroprotection within the dose range of 0.1–1 mg/kg. However, higher doses of L-NNA (3 and 10 mg/kg i.p.), which caused an almost complete abolition of NOS activity, failed to significantly reduce infarct size. It now seems that an incomplete, partial inhibition of NOS activity in the brain may actually have greater efficacy for the treatment of permanent focal cerebral ischemia than does a complete or more extensive inhibition.

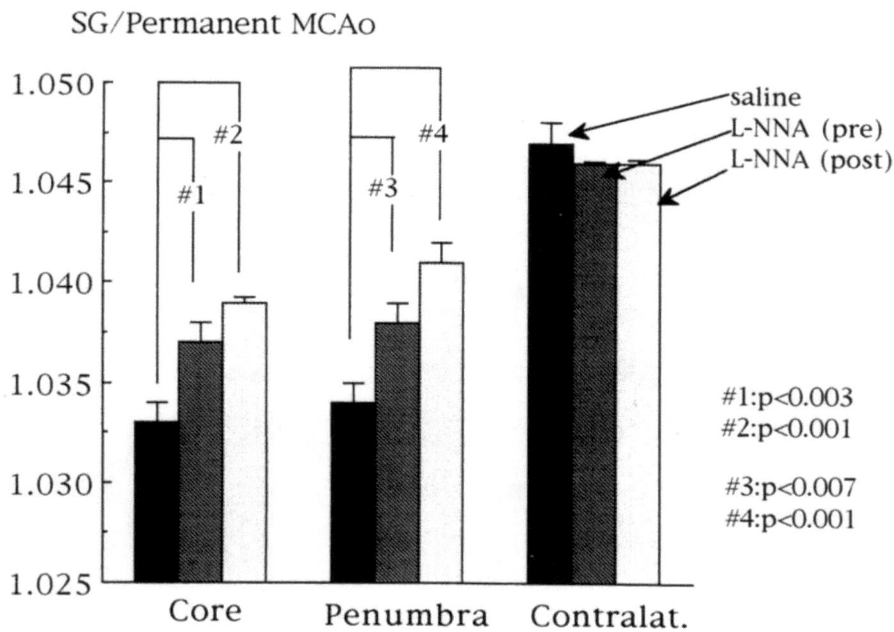

Fig. 3. Effects of L-NNA (1 mg/kg, i.p.) on ischemic brain edema (cortical specific gravity, *SG*). Ischemic insult was caused by 4h MCA occlusion in rats. Saline (*black bars*), 1 ml/kg; L-NNA (pre, *shaded bars*), 1 mg/kg i.p., 5 min before and 3h after MCA occlusion; L-NNA (post, *light bars*), 1 mg/kg i.p. 5 min and 3h after MCA occlusion. Each bar represents the mean ± SEM (ANOVA, followed by Dunnett test)

Difference in the Timing of Drug Administration

An alternative possible explanation for the discrepancies reported on the effects of NOS inhibitors in focal cerebral ischemia might be the different time intervals of NOS inhibition following ischemia. In many of the reports of positive amelioration, NOS inhibitors were repeatedly administered over a postischemic period of 1 day, while in those reporting no amelioration, only acute administration was employed in the pre- and/or postischemic period. It is conceivable that cellular events associated with NOS in the early as compared to later periods following cerebral ischemia may be different. Because i-NOS expression requires at least 3–4 h [78–81], this may be one good reason to consider the role of i-NOS in nonneuronal cells during the onset and development of ischemic injury.

Differences in the Types of NOS Inhibitors

L-NMMA blocks basal but not agonist-induced production of NO within the rat aorta, while L-NNA blocks both [82], and L-NMMA has no effect on $O_2\cdot^-$ generation from purified brain NOS while L-NAME inhibits this radical production [83]. However, positive and negative ameliorative effects have been shown with both L-NNA [68,69,75,76] and L-NAME [66,67,73,74,77,84–86] in animals subjected to cerebral ischemia, but the effects of L-NMMA on ischemic brain damage has not yet been studied.

Janus Faces of NO [87,88]

Lipton and his colleagues [89,90] have elegantly shown that the neuroprotective and neurodestructive effects of NOS inhibitors or nitroso compounds are dependent on their oxidation-reduction state. In that report, it was demonstrated that NO·-mediated neurotoxicity is generated, at least in part, by reaction with superoxide anion ($O_2\cdot^-$), apparently leading to the formation of peroxynitrite ($ONOO^-$), and not by NO· alone. In contrast, the neuroprotective effects of NO result from downregulation of NMDA receptor activity by reaction with thiol group(s) at the receptor redox modulatory site. This reaction is not mediated by NO· itself, but occurs under conditions supporting *S*-nitrosylation of the NMDA receptor thiol. The Essentially same observations were also demonstrated by Manzoni et al. [91]. These results suggest that a feedback inhibition of NMDA receptors by NO may be operating under physiological conditions. Furthermore, Manzoni et al. noted that the redox versatility of NO allows for its conversion from a neuroprotective to a neurotoxic species by a change in the ambient redox milieu. Because these observations seem to be highly significant, this hypothesis must be tested in vivo to see if it can resolve the paradoxical effects of NOS inhibitors on rodent models of cerebral ischemia.

The Neuroprotective Effect of NOS Inhibition in Transient Focal Cerebral Ischemia

As described next, study of direct measure of cerebral NO content demonstrates that a transient increase of NO was observed in the acute phase of permanent MCA occlusion in rats and that CBF restoration is likely to enhance NO production and contribute NO to the postischemic tissue injury. Thus, it may be suggested that the neuroprotective effects of the NOS inhibitor would be more enhanced in transient than in permanent ischemia.

We have demonstrated that L-NNA (1 mg/kg, 5 min post ischemia and 30 min of reperfusion, i.p.) attenuated the decrease in cortical SG of rats subjected to 2 h of reperfusion after 2 h MCAo, using the miniclip (Fig. 4). Ashwal et al. [85] and Matsui et al. [92] have also shown that ameliorative effects were obtained by an infusion of low-dose L-NAME, which caused MABP to remain unaltered during the experimental period. They administered a 0.1-mg/kg bolus followed by 0.01 mg kg−1 min−1 of L-NAME during 1 h before occlusion, 3 h of MCA occlusion, and 2 h of reperfusion.

Dawson et al. [84] developed transient MCA occlusion induced by application of endothelin-I to the MCA. By treating the normotensive rats subjected to this model with L-NAME, they have not seen ameliorative results, but a dramatic reduction in lesion size (55%) has been reported after 3 h of MCA occlusion and 2 h of reperfusion in the SHR [85]. L-NAME (15 mg/kg i.v.) administered 15 min before a 2-h MCA occlusion induced by the intraluminal filament technique has been reported to significantly increase lesion volume after 3 days of reperfusion [74]. Because the intraluminal filament technique is likely to cause endothelial damage, leading to microcirculatory disturbance, this high dose of L-NNA might prevent restoration of the normal level of blood flow after recirculation. Alternatively, L-NAME might maintain the elevated level of glutamate after recirculation. The neuroprotective efficacy of NO synthase inhibitors may be critically dependent on the dose of drug administered and the difference in ischemic model.

Effect of NO Donor in Focal Cerebral Ischemia

Several investigators have attempted to examine the effects of NO production on the ischemic brain, as an activation of the NO-cGMP (cyclic guanosine monophosphate) pathway will be expected to increase CBF led by vasodilation, inhibition of platelet aggregation, or neutrophil adhesion to endothelium. The administration of L-arginine, but not D-arginine, reduced infarct size after MCA occlusion in SHR and Sprague-Dawley (SD) rats [93,94]. NO donors SNP and 3-morpholino-sydnonimine (SIN-1) increase rCBF in the ischemic region and reduce tissue damage in both rats and SHR [86,95]. These results may be summarized as follows.

1. The moderate CBF increase in the ischemic region (by 50%) may be sufficient to raise blood flow above the ischemic threshold and therefore salvage tissue in the penumbral regions [86]. This phenomenon may be ascribable to NO-induced vasodilation.

2. Furthermore, it should be considered in the sense that NO-induced inhibition of platelet aggregation and adhesion may improve blood flow in the microvasculature. Although infusion of SNP does not affect platelet aggregation in nonischemic, normotensive rats [94], platelet activation is likely to be greater in ischemic versus nonischemic tissue and may be more pronounced in SHR rats compared with normotensive strains. Thus, the antiplatelet properties of NO donors may contribute to their neuroprotective action in the context of focal cerebral ischemia.

3. Within the first hour after MCA occlusion, extracellular glutamate in the cerebral cortex reaches its peak and subsequently declines to reach a steady state 2 h post occlusion. However, SNP or SIN-1 can inhibit glutamate binding to the NMDA recep-

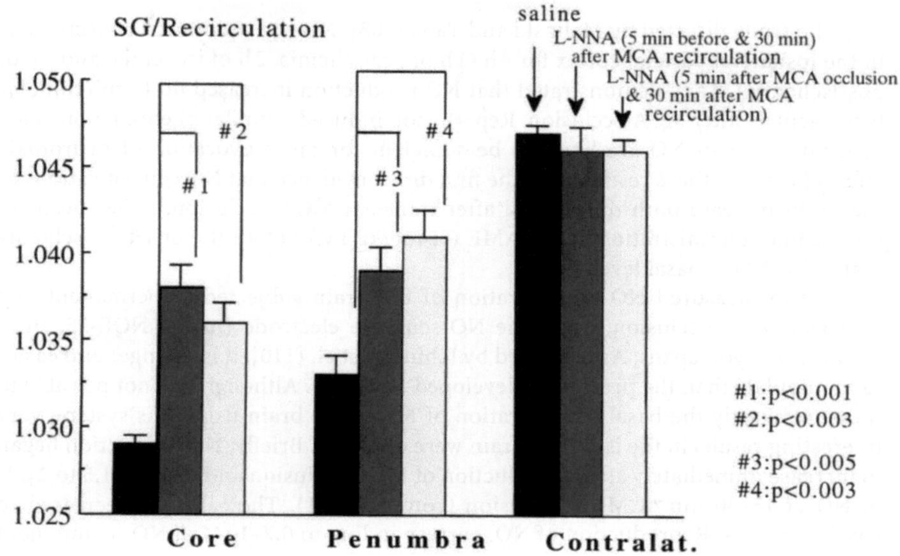

Fig. 4. Effects of L-NNA on ischemic brain edema (cortical specific gravity, *SG*). Ischemic insult was caused by 2-h reperfusion following 2-h MCA occlusion in rats. Saline, 1 ml/kg; L-NNA (pre), 1 mg/kg i.p., 5 min before and 30 min after MCA recirculation; L-NNA (post), 1 mg/kg i.p., 5 min after MCA occlusion and 30 min after MCA recirculation. Each bar represents the mean ± SEM (ANOVA, followed by Dunnett test)

tor site by an NO-independent action [95], or, NO produced from any source may reduce Ca^{2+} influx through the NMDA receptor channel by nitrosylation and subsequent oxidation of the redox modulatory site.

4. The effect of L-arginine is known to be sustained for at least 90 min after cessation of L-arginine infusion [93], perhaps because of sequestration of L-arginine into endothelial cells, allowing continued production of NO and enhancement of blood flow beyond the duration of the infusion.

Here the following question is raised: Is it likely that an extrinsic L-arginine really contributes to NO production in the ischemic brain? Taken together with the fact that a transient increase in blood flow during the first hour of permanent MCA occlusion has a beneficial effect which is still apparent 24 h later, other mechanism(s) may be involved.

Nitric Oxide Production in Brain Subjected to Permanent and Transient Focal Cerebral Ischemia

Many methods to detect NO or its degradative products in biological solutions have been recently made available, including methods that use quantitative oxidation of oxyhemoglobin [96,97], metal complexes [98], Saltzman reagents [99], Griess reagents [100], gas chromatography [101], chemiluminescence [102,103], electron paramagnetic resonance (EPR) [104], electron spin resonance (ESR) [105], and the NO-sensitive microelectrode [106–110].

In the study directed by Malinski and Taha [108], NO was monitored continuously in the ipsilateral parietal cortex for 4 h (1 h of preischemia, 2 h of ischemia, and 1 h of postischemia). They demonstrated that NO production increased up to micromolar levels acutely after MCA occlusion. Reperfusion induced a similar phenomenon. Such concentrations of NO are likely to be sufficient for the provocation of neurotoxic effects [90], and these results were the first direct evidence that NO concentrations in the brain increase both during and after transient MCA occlusion. It has been reported that administration of L-NAME (30 mg/kg i.v.) before the onset of ischemia maintains NO at basal levels.

We also measured NO concentration of the brain subjected to permanent and transient MCA occlusion, using the NO-sensitive electrode (model NOE-55, Inter Medical, Tokyo, Japan). As described by Ichimori et al. [110], it is stronger and easier to manipulate than the previously developed electrode. Although it is not possible to measure exactly the basal concentration of NO in the brain using this system, very interesting results in the ischemic brain were obtained. Briefly, NO production began to increase immediately after an induction of MCA occlusion and reached 1.5 to 2 μM of NO at 15–30 min of MCA occlusion (component 1). Thereafter, it decreased to baseline values. Reproduction of NO, corresponding to 0.7–1 μM of NO (component

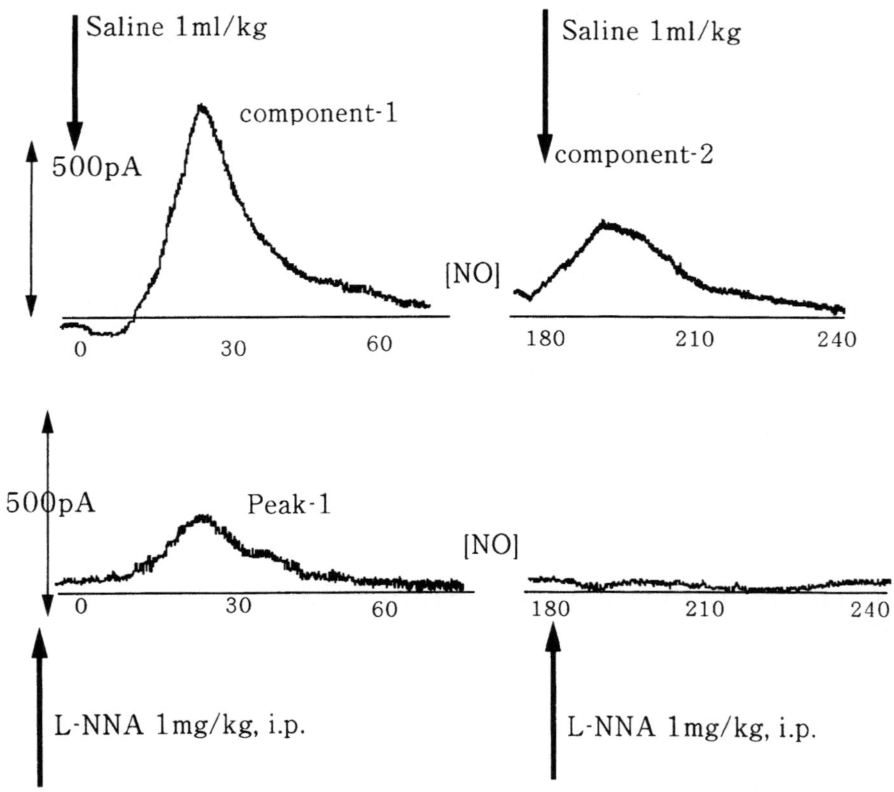

Fig. 5. Temporal feature of cerebral nitric oxide concentration [*NO*] in permanent MCA occlusion in the rat

2), was monitored 3–4 h after MCA occlusion. Administration of L-NNA (1 mg/kg i.p.) at 5 min before and 3 h after onset of MCA occlusion significantly reduced both NO production (component 1 by half and component 2 almost completely) and simultaneously mitigated brain infarction (Fig. 5). Reperfusion following 2 h of MCA occlusion led to an increase in NO production to 1–1.5μM at 1–2 h (component 2′). Administration of L-NNA (1 mg/kg i.p.) 5 min before the induction of MCA occlusion and 1 h after reflow not only reduced NO production (component 2′, almost completely) during these periods but also diminished the subsequent cerebral infarct. We first reported this biphasic increase during the acute phase of permanent and transient MCA occlusion. Although these two groups have developed well the polarographical method for NO measurement, they need to investigate whether the electrode is absolutely specific to NO in the in vivo condition [111].

Tominaga et al. [112] measured NO spin adducts using diethyldithiocarbamate and Fe as spin-trapping agents by cryogenic EPR. They showed that an increased amount of spin adducts of NO radicals in the brain was detectable at 5–30 min after the onset of focal cerebral ischemia. Kader et al. [113] elucidated a sharp transient increase in the activity of NOS during the first 60 min of cerebral ischemia in rats. In parallel, the content of cGMP and nitrite increased in the affected hemisphere. Kumura et al. [114] detected nitrite (NO_2^-) and nitrate (NO_3^-) as end products of NO in plasma during acute focal cerebral ischemia and reperfusion in rats.

These direct and indirect methods provide ample evidence for the possibility that NO production may be transiently or biphasically enhanced in the early phase of focal cerebral ischemia and that the amount of NO production may sufficiently contribute to ischemic brain injury. However, here are several problems to be considered. Although it is acceptable that NO production is enhanced in brain subjected to permanent and transient focal ischemia, what is the possible site of NO source? Which type(s) are NOS is responsible for this phenomenon? The next section focuses on these questions.

Time-Course of NOS Activity of Cerebral Microvessels and Cerebral Cortex Taken from Brain Subjected to Permanent and Transient MCA Occlusion in Rats

It is well known that the microvascular endothelium, including the endothelial cells of intraparenchymal arteries, venules, and capillary networks, are the major constituents of the blood–brain barrier regulating the permeability of intrinsic and extrinsic substances [115]. The fact that NOS inhibitors were beneficial in the treatment of ischemia-induced [66–69] and cold injury-induced vasogenic brain edema [72] encouraged us to investigate the temporal alteration in NOS activity of cerebral microvessels taken from an in vivo focal cerebral ischemic model.

The significant elevation of Ca^{2+}-dependent c-NOS activity in brain microvessels isolated from the affected hemisphere at 15 min after MCA occlusion returned to the control level at 4 h in rats (Fig. 6). Ca^{2+}-independent i-NOS activity increased ninefold at 4 h and 24 h after MCA occlusion, respectively [56]. Two major conclusions can be drawn from these results. First, the cerebral microvessels possess two distinct types of NOS, i.e., Ca^{2+}-dependent c-NOS and Ca^{2+}-independent i-NOS. This finding confirms an earlier study that employed endothelial cells cultured from mouse brain [116]. Second, NOS activity in brain microvessels is enhanced following cerebral ischemia.

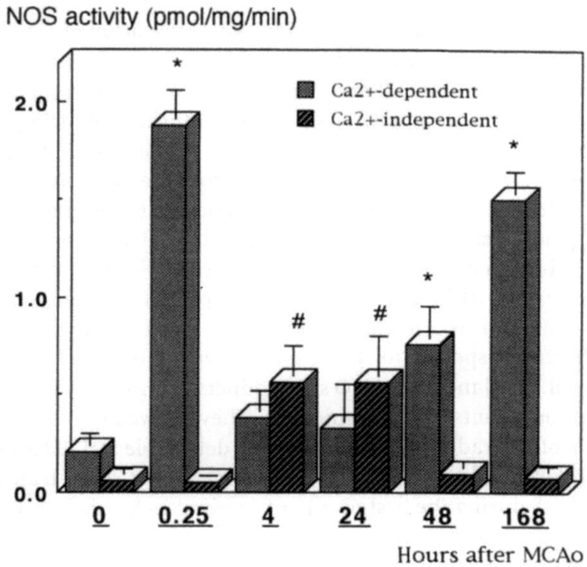

Fig. 6. Temporal alterations in Ca^{2+}/calmodulin-dependent (*shaded bars*) and -independent (*hatched bars*) nitric acid synthase (NOS) activities in brain microvessels isolated from the affected hemisphere after left MCA (*MCAo*) occlusion in rats. Following MCA occlusion, animals were killed by transcardial perfusion with saline containing 10 IU/ml of heparin under anesthesia with ether at the times indicated on the abscissa. Brain microvessels from the affected left hemisphere were isolated as previously reported [54]. NOS activity was measured by monitoring the conversion from L-[^{3}H]arginine to L-[^{3}H]citrulline. Ca^{2+}/calmodulin-dependent NOS activity was calculated by subtracting the enzyme activity observed in Ca^{2+}/calmodulin-free medium from that observed in Ca^{2+}/calmodulin-containing medium. Ca^{2+}/calmodulin-independent NOS activity was determined in Ca^{2+}/calmodulin-free medium [54]. Each histogram represents the mean ± SEM. *, $P < 0.05$ vs. naive control (Ca^{2+}/calmodulin-dependent NOS); #, $P < 0.05$ vs. naive control (Ca^{2+}/calmodulin-independent NOS) (ANOVA, followed by Dunnett test)

This is the first demonstration to our knowledge that NOS activation in definite cell types, i.e., brain microvessels, may be responsible for the increased production of NO subsequent to cerebral ischemia.

With the same method, NOS activity was measured in cerebral microvessels obtained from brain subjected to reperfusion. A 2-h reperfusion after a 2-h MCA occlusion led to enormous activation of Ca^{2+}-dependent c-NOS (Fig. 7).

We next measured the time-course of NOS activity in soluble (S2) and insoluble fractions (17 000 × g × 20 min) obtained from rat cerebral cortex subjected to permanent and transient focal cerebral ischemia. The S2 fraction is thought to contain N-c-NOS. Briefly, during the period observed (4 h after onset of MCA occlusion), activity of Ca^{2+}-independent NOS (i-NOS) was not detected. However, that of Ca^{2+}-dependent NOS (c-NOS), presumably corresponding to N-c-NOS, increased to twofold of baseline value (unpublished data).

The temporal alterations in E-c-NOS and N-c-NOS activity following MCA occlusion in rats are strongly correlated with the fact that both enzymes are upregulated in the acute phase of cerebral ischemia [117,118].

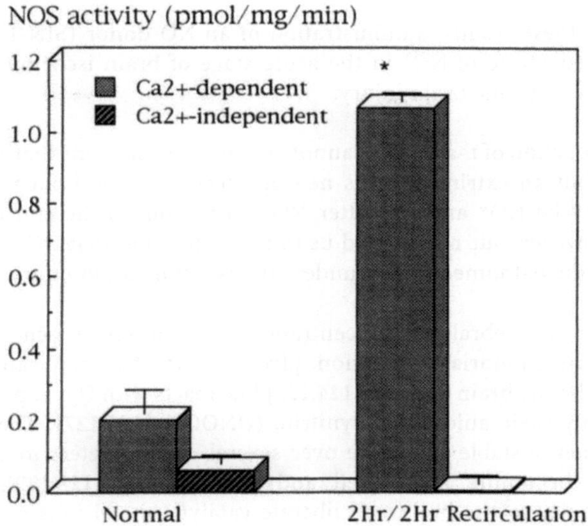

Fig. 7. A specific enhancement of Ca^{2+}/calmodulin-dependent NOS activity in brain microvessels obtained at 2-h recirculation following 2-h MCA occlusion in rats. Isolation of brain microvessels from the ischemically affected left hemisphere and NOS activity assay were carried out as described for Fig. 6. Each histogram represents the mean ± SEM for the number of animals shown in parentheses. *, $P < 0.05$ vs. naive control (Ca^{2+}/calmodulin-dependent NOS) (ANOVA, followed by Dunnett test)

Possible Mechanisms of Nitric Oxide-Mediated Ischemic Brain Injury

Although the free radical form of NO can directly induce DNA damage when present in high concentrations (>5 m*M*) [119,120], in the studies on NO measured in vivo, including our results described here, the relatively low concentrations of NO (e.g., 2–3 μ*M*) after permanent focal ischemia [121] and the short biological half-life of NO [2,49] suggest that NO may not directly mediate excitotoxic damage. Experimental evidence thus far collected is as follows:

1. NOS blockade by a low dose of NOS inhibitor exacerbates ischemic neuronal damage.
2. Cerebral ischemia leads to transient NO production, ranging from 2 to 3 μ*M* in brain tissue.
3. NO production may be ascribed to activation of N-c-NOS (cortex) and E-c-NOS (cerebral microvessels) in the early stages of ischemia and recirculation. The i-NOS activity of cerebral microvessels appeared prominent at the later phase (4–24 h).
4. An overproduction of NO via activation of N-c-NOS is mainly responsible for neuronal damage following permanent focal cerebral ischemia [122].
5. There have been no precise data on cerebral NO concentrations in the early stage of recirculation.

6. Contrary to these results, administration of an NO donor (SIN-1 or SNP) or L-arginine (a substrate of NO) in the acute stage of brain ischemia was found to attenuate the ischemic brain injury.

Although the action of L-arginine cannot be explained, it seems that an intrinsic NO is neurotoxic but an extrinsic NO is neuroprotective in the ischemic brain. If so, the former may be NO* and the latter NO^+, according to the concept of Stamler et al. [123]. However, our results lead us to the conclusion that NO acts as a neurotoxin and that the pathomechanism underlying ischemic brain injury is explained as follows.

The increase in cerebral NO concentration at the onset of ischemia causes the inhibition of mitochondrial respiration, glycosis, and DNA replication, leading to initiation of ischemic brain damage [124,125] like reacts with O_2^- at physiological pH to yield a highly toxic anion, peroxynitrite ($ONOO^-$) [126,127]. The peroxynitrite anion is sufficiently stable to diffuse over several cell diameters to critical cellular targets before becoming protonated and decomposing [128,129]. Ischemia/reoxygenation injury to the brain can liberate catalytic metal ions and increase the formation of reactive radicals. The peroxynitrite anion is a potent activator of lipid peroxidation [130]; it is protonated to peroxynitrous acid (ONOOH), and at the acidotic condition, it easily decays to form highly reactive free radical species, hydroxyl radical (OH·) and nitrogen dioxide (NO_2^-) [126,131,132]. Decomposition of NO and peroxinitrite anion to form hydroxyl radicals better explains the result that NO concentration in the ischemic brain returns to the baseline value during the first 2–3 h of MCA occlusion (see Fig. 5), despite the enhancement of c-NOS activity of cerebral cortex and microvessels.

The NMDA receptor, which is activated by an increase in extracellular glutamate following cerebral ischemia, stimulates arachidonic acid release [133], and NO directly interacts with cyclo-oxygenase (COX) to increase enzymatic activity in neurons [134]. Brain microvessels also have a potent activity to generate COX and lipoxygenase (LPX) products. Neurons and brain microvessels are liable to be exposed to the superoxide anion via an activation of COX, leading to formation of the peroxynitrite anion. This anion may acts as a neurotoxin, as has been described.

Furthermore, NO facilitates neurotransmitter release or inhibits its uptake in neurons including dopamine [135], acethylcholine [136], serotonin, and L-glutamate [137] in the central nervous system. The facilitation of neurotransmitter release induced by NO might be elicited via an inhibition of neuronal Na^+/K^+-ATPase, leading to an accumulation of L-glutamate in the extracellular space. A resultant excessive glutamate accumulation in the extracellular space causes overstimulation of NMDA receptors in the postsynapse, leading to an activation of c-NOS through Ca^{2+} influx and a malignant derangement of NO production.

In contrast, this enzyme of brain microvessels is activated by a lipid peroxide of arachidonic acid, which cellular event is thought to contribute to edema formation [138]. Free radical species such as NO·, O_2^-, $ONOO^-$, OH·, and NO_2^- might induce an enhancement of Na^+/K^+-ATPase as well as membrance perturbation and increased permeability in brain microvessels. These cellular events participate in the mechanisms of edema formation.

It is hypothesized that NO synthesized from cerebral microvascular endothelium may diffuse out and act as a killer for neighboring neurons. Recent histological examinations disclosed that NOS neurons lie near branching points of the arteriole that descends through the cerebral cortex from its pial surface, and they are close to intracerebral blood vessels [139]. In addition, E-c-NOS is predominantly membrane bound [103]; however, when an agonist-induced Ca^{2+} influx acts on the membrane of microvessels that contact the circulating blood, E-c-NOS is phosphorylated [46]. This spatial relationship explains the neuronal control of cortical blood flow by NOS neurons [140,141], possibly mediated by the rapid action of NO. The condition of permanent and transient cerebral ischemia highly stimulates these interactions over the physiological range. Not only N-c-NOS but also E-c-NOS may be deeply involved in ischemic brain injury. The fact that E-c-NOS is greatly activated after MCA recirculation may support the hypothesis that E-c-NOS activation contributes more to cerebral injury than permanent ischemia because of reperfusion.

Finally, we should comment on the activation of i-NOS of cerebral microvessels at 4–24h after MCA occlusion. Because the temporal threshold for cerebral infarction is thought to be 2–3h in the area with severe ischemia, i-NOS may contribute to the development of ischemic brain injury but is not essential.

Although we have never examined the role of NO donors in cerebral ischemia, ischemia may be brought about by the nitrosylation of the NMDA receptor by NO^+ as described by Stamler et al. It is deeply impressive that NO acts as an opposite functional mediator in cerebral ischemia. We believe that this chapter contributes to the ongoing study and will open up a new therapuetical strategy for cerebrovascular disease.

References

1. Dawson VL, Dawson TM, Bartley DA, Uhl GR, Snyder SH (1993) Mechanisms of nitric oxide-mediated neurotoxicity in primary brain cultures. J Neurosci 13:2651–2661
2. Moncada S, Palmer RMJ, Higgs EA (1991) Nitric oxide: physiology, pathophysiology, and pharmacology. Pharmacol Rev 43:109
3. Dawson TM, Bredt DS, Fotuhi M, Hwang PM, Snyder SH (1991) Nitric oxide synthase and neuronal NADPH diaphorase are identical in brain and peripheral tissues. Proc Natl Acad Sci USA 88:7797–7801
4. Knowles RG, Palacious M, Palmer RMJ, Moncada S (1989) Formation of nitric oxide from L-arginine in the central nervous system: a transduction mechanism for stimulation of the soluble guanylate cyclase. Proc Natl Acad Sci USA 86:5159–5162
5. Bredt DS, Snyder SH (1990) Isolation of nitric oxide synthetase, a calmodulin-requiring enzyme. Proc Natl Acad Sci USA 87:682–685
6. Bredt DS, Hwang PM, Snyder SH (1990) Localization of nitric oxide synthase—indicating a neuronal role for nitric oxide. Nature 347:768–770
7. Bredt DS, Hwang PM, Glatt CE, Lowenstein C, Reed RR, Snyder SH (1991) Cloned and expressed nitric oxide synthase structurally resembles cytochrome P-450 reductase. Nature 351:714–718
8. Lamas S, Marsden PA, Li GK, Tempst P, Michel T (1992) Endothelial nitric oxide synthase: molecular cloning and characterization of a distinct constitutive enzyme isoform. Proc Natl Acad Sci USA 89:6348–6352
9. Lowenstein CJ, Snyder SH (1992) Nitric oxide, a novel biologic messenger. Cell 70:705–707
10. Lowenstein CJ, Alley EW, Raval P, Snowman AM, Snyder SH, Russel SW, Murphy WJ (1993) Macrophage nitric oxide synthase gene: two upstream regions mediate induction by interferon g and lipopolysaccharide. Proc Natl Acad Sci USA 90:9730–9734
11. Lyons CR, Orloff GJ, Cunningham JM (1992) Molecular cloning and functional expression of an inducible nitric oxide synthase from a murine macrophage cell line. J Biol Chem 267:6370–6374

12. Marsden PA, Schappert KT, Chen HS, Flowers M, Sundell CL, Wilcox JN, Lamas S, Michel T (1992) Molecular cloning and characterization of human endothelial nitric oxide synthase. FEBS Lett 307:287–293
13. Nishida K, Harrison DG, Navas JP, Fisher AA, Dockery SP, Uematsu M, Nerem RM, Alexander RW, Murphy TJ (1992) Molecular cloning and characterization of the constitutive bovine aortic endothelial cell nitric oxide synthase. J Clin Invest 90:2092–2096
14. Xie Q-W, Cho HJ, Calaycay J, Mumford RA, Swiderek KM, Lee TD, Ding A, Troso T, Nathan C (1992) Cloning and characterization of inducible nitric oxide synthase from mouse macrophages. Science 256:225–228
15. Charles IG, Chubb A, Gill R, Clare J, Lowe PN, Holmes LS, Page M, Keeling JG, Moncada S, Riveros-Moreno V (1993) Cloning and expression of a rat neuronal nitric oxide synthase coding sequnence: a baculovirus/insect cell system. Biochem Biophys Res Commun 196:1481–1489
16. Charles IG, Palmer RMJ, Hickery MS, Bayliss MT, Chubb AP, Hall VS, Moss DW, Moncada S (1993) Cloning, characterization, and expression of a cDNA encoding an inducible nitric oxide synthase from the human chondrocyte. Proc Natal Acad Sci USA 90:11419–11423
17. Geller DA, Lowenstein CJ, Shapiro RA, Nussler AK, Silvio MD, Wang SC, Nakayama DK, Simmons RL, Snyder SH, Billiar TR (1993) Molecular cloning and expression of inducible nitric oxide synthase from human hepatocytes. Proc Natl Acad Sci USA 90:3491–3495
18. MacNaul KL, Hutchinson NI (1993) Differential expression of iNOS and cNOS mRNA in human vascular smooth muscle cells and endothelial cells under normal and inflammatory conditions. Biochem Biophys Res Commun 196:1330–1334
19. Nunokawa Y, Ishida N, Tanaka S (1993) Cloning of inducible nitric oxide synthase in rat vascular smooth muscle cells. Biochem Biophys Res Commun 191:89–94
20. Sherman PA, Laubach VE, Reep BR, Wood ER (1993) Purification and cDNA sequence of an inducible nitric oxide synthase from a human tumor cell line. Biochemistry 32:11600–11605
21. Wood ER, Berger H Jr, Sherman PA, Lapetina EG (1993) Hepatocytes and macrophages express an identical cytokine inducible nitric oxide synthase gene. Biochem Biophys Res Commun 191:767–774
22. Chartrain NA, Geller DA, Koty PP, Sitrin NF, Nussler AK, Hoffman EP, Billiar TR, Hutchinson NI, Mudgett JS (1994) Molecular cloning, structure, and chromosomal localization of the human inducible nitric oxide synthase gene. J Biol Chem 269:6765–6772
23. Nathan C (1992) Nitric oxide as a secretory product of mammalian cells. FASEB J 6:3051–3064
24. Marletta MA (1993) Nitric oxide synthase structure and mechanism. J Biol Chem 268:12231–12234
25. Mayer B, John M, Böhme E (1990) Purification of a Ca^{2+}/calmodulin-dependent nitric oxide synthase from porcine cerebellum: cofactor role of tetrahydrobiopterin. FEBS Lett 277:215–219
26. Schmidt HHW, Pollock JS, Nakane M, Gorsky LD, Förstermann U, Murad F (1991) Purification of Ca^{2+}/calmodulin-dependent nitric oxide synthase from porcine cerebellum. Proc Natl Acad Sci USA 88:365–369
27. Hiki K, Hattori R, Kawai C, Yui Y (1992) Purification of insoluble nitric oxide synthase from rat cerebellum. J Biochem 111:556–558
28. Ohshima H, Oguchi S, Adachi S, Suzuki H, Sugimura T, Esumi H (1992) Purification of nitric oxide synthase from bovine brain: immunological characterization and tissue distribution. Biochem Biophys Res Commun 183:238–244
29. Förstermann U, Pollock JS, Schmidt HHHW, Heller M, Murad F (1991) Calmodulin-dependent endothelium-derived relaxing factor/nitric oxide synthase activity is present in the particulate and cytosolic fractions of bovine aortic endothelial cells. Proc Natl Acad Sci USA 88:1788–1792

30. **Förstermann U, Schmidt HHHW, Pollock JS, Sheng H, Mitchell JA, Warner TD, Nakane M, Murad F (1991) Isoforms of nitric oxide synthase: characterization and purification from different cell types. Biochem Pharmacol 42:1849–1857**
31. **Busconi L, Michel T (1993) Endothelial nitric oxide synthase. J Biol Chem 268:8410–8413**
32. **Nakane M, Schmidt HHHW, Pollock JS, Förstermann U, Murad F (1993) Cloned human brain nitric oxide synthase is highly expressed in skeletal muscle. FEBS Lett 316:175–180**
33. **Murphy S, Minor RL, Welk G Jr, Harrison DG (1990) Evidence for an astrocyte-derived vasorelaxing factor with properties similar to nitric oxide. J Neurochem 55:349–351**
34. Murphy S, Simmons ML, Agullo L, Garcia A, Feinstein DL, Galea E, Reis DJ, Minc-Golomb D, Schwartz JP (1993) Synthesis of nitric oxide in CNS glial cells. Trends Neurosci 16:323–328
35. Galea E, Feinsten DL, Reis DJ (1992) Induction of calcium-independent nitric oxide synthase activity in primary rat glial cultures. Proc Natl Acad Sci USA 89:10945–10949
36. Baek KJ, Thiel BA, Lucas S, Stuehr DJ (1993) Macrophage nitric oxide synthase subunits: purification, characterization, and prosthetic groups and substrate in regulation of their association into a dimeric enzyme. J Biol Chem 268:21120–21129
37. Sherman M, Aeberhard EE, Wong VZ, Griscavage JM, Ignaro LJ (1993) Pyrrolidine dithiocarbamate inhibits induction of nitric oxide synthase activity in rat alveolar macrophages. Biochem Biophys Res Commun 191:1301–1308
38. Smith JA, Radomski MW, Schulz R, Moncada S, lewis MJ (1993) Porcine ventricular endocardial cells in culture express the inducible form of nitric oxide synthase. Br J Pharmacol 108:1107–1110
39. Hecker M, Mülsch A, Busse R (1994) Subcellular localization and characterization of neuronal nitric oxide synthase. J Neurochem 62:1524–1529
40. Dawson VL, Dawson TM, London ED, Bredt DS, Snyder SH (1991) Nitric oxide mediates glutamate neurotoxicity in primary cortical cultures. Proc Natl Acad Sci USA 88:6368–6371
41. Loiacono RE, Beart PM (1992) Hippocampal lesions induced by microinjection of the nitric oxide donor nitroprusside. Eur J Pharmacol 216:331–333
42. Choi DW (1988) Calcium-mediated neurotoxicity: relationship to specific channel types and role in ischemic damage. Trends Neurosci 11:465–469
43. Hewett SJ, Corbett JA, McDaniel ML, Choi DW (1993) Inhibition of nitric oxide formation dose not protect murine cortical cell cultures from NMDA neurotoxicity. Brain Res 625:333–341
44. Rees DD, Palmer RMJ, Schulz R, Hodson HF, Moncada S (1990) Characterization of three inhibitors of endothelial nitric oxide synthase in vitro and in vivo. Br J Pharmacol 101:746–752
45. McCall TB, Feelisch M, Palmer RMJ, Moncada S (1991) Identification of *N*-iminoethyl-L-ornithine as an irreversible inhibitor of nitric oxide synthase in pagocytic cells. Br J Pharmacol 102:234–238
46. Michel AD, Phul RK, Stewart TL, Humphrey PA (1993) Characterization of the binding of [^{3}H]-L-N^G-nitro-arginine in rat brain. Br J Pharmacol 109:287–288
47. Klatt P, Schmidt K, Brunner F, Mayer B (1994) Inhibitors of brain nitric oxide synthase: binding kinetics, metabolism, and enzyme inactication. J Biochem 269:1674–1680
48. Tabrizi-Fard MA, Fung HL (1994) Pharmacokinetics, plasma protein binding and urinary excretion of N^ω-nitro-L-arginine in rat. Br J Pharmacol 111:394–396
49. Furfine ES, Harmon MF, Paith JE, Garrey EP (1993) Selective inhibition of constitutive nitric oxide synthase by L-N^G-nitroarginine. Biochemistry 32:8512–8517
50. Lambert LE, French JF, Whitten JP, Baron BM, McDonald IA (1992) Characterization of cell selectivity of two novel inhibitors of nitric oxide synthesis. Eur J Pharmacol 216:131–134

51. Mayer B, Schmid M, Klatt P, Schmidt K (1993) Reversible inactivation of endothelial nitric oxide synthase by N^G-nitro-L-arginine. FEBS Lett 333:203–206
52. Olken NM, Marletta MA (1992) N^G-Allyl- and N^G-cyclopropyl-L-arginine; two novel inhibitors of macrophage nitric oxide synthase. J Med Chem 35:1137–1144
53. Dwyer MA, Bredt DS, Snyder SH (1991) Nitric oxide synthase: irreversible inhibition by L-N^G-nitroarginine in brain in vitro and in vivo. Biochem Biophys Res Commun 176:1136–1141
54. Hasan K, Heesen B-J, Corbett JA, McDaniel ML, Chang K, Allison W, Wolffenbuttel BHR, Williamson JR, Tilton RG (1993) Inhibition of nitric oxide formation by guanidines. Eur J Pharmacol 249:101–106
55. Griffiths MJD, Messent M, MacAllister RJ, Evans TW (1993) Aminoguanidine selectively inhibits inducible nitric oxide synthase. Br J Pharmacol 110:963–968
56. Nagafuji T, Sugiyama M, Matsui T (1994) Temporal profiles of Ca^{2+}/calmodulin-dependent and -independent nitric oxide synthase activity in the rat brain microvessels following cerebral ischemia. In: Ito U (eds) Brain edema IX. Acta Neurochir 60:285–288
57. Salter M, Knowles RG, Moncada S (1991) Widespread tissue distribution, species distribution and changes in activity of Ca^{2+}-dependent and Ca^{2+}-independent nitric oxide synthase. FEBS Lett 291:145–149
58. Moore PK, Babbedge RC, Wallace P, Gaffen ZA, Hart SL (1993) 7-Nitro indazole, an inhibitor of nitric oxide synthase, exhibits anti-nociceptive activity in the mouse without increasing blood pressure. Br J Pharmacol 108:296–297
59. Moore PK, Wallace P, Gaffen Z, Hart SL, Babbedge RC (1993) Characterization of the novel nitric oxide synthase inhibtor 7-nitro indazole and related indazoles; antinociceptive and cardiovascular effects. Br J Pharmacol 110:219–224
60. Babedge RC, Bland-Ward PA, Hart SL, Moore PK (1993) Inhibition of rat cerebellar nitric oxide synthase by 7-nitro indazole and related substituted indazoles. Br J Pharmacol 110:225–228
61. Medhurst AD, Greenless C, Parsons AA, Smith SJ (1994) Nitric oxide synthase inhibitors 7- and 6-nitroindazole relax smooth muscle in vitro. Eur J Pharmacol 256:R5–R6
62. Zingarelli B, Carnuccio R, Rosa MD (1993) Cloricromene inhibits the induction of nitric oxide synthase. Eur J Pharmacol 243:107–111
63. Buchmüller-Rouiller Y, Schneider P, Betz-Corradin S, Smith J, Mauël J (1992) 3-Amino-1,2,4-triazole inhibits macrophage NO synthase. Biochem Biophys Res Commun 183:150–155
64. Palacios M, Padron J, Glaria L, Rojas A, Delgado R, Knowles R, Moncada S (1993) Chlorpromazine inhibits both the constitutive nitric oxide synthase and the induction of nitric oxide synthase after LPS challenge. Biochem Biophys Res Commun 196:280–286
65. Szabo C, Thiemermann C, Vane JR (1993) Dihydropyridine antagonists and agonists of calcium channels inhibit the induction of nitric oxide synthase by endotoxin in cultured macrophage. Biochem Biophys Res Commun 196:825–830
66. Buisson A, Margaill I, Callebert J, Plotkine M, Boulu RG (1993) Mechanisms involved in the neuroprotective activity of a nitric oxide synthase inhibitor during focal cerebral ischemia. J Neurochem 61:690–696
67. Buisson A, Plotkine M, Boulu RG (1992) The neuroprotective effect of a nitric oxide inhibitor in a rat model of focal cerebral ischemia. Br J Pharmacol 106:766
68. Nowicki JP, Duval D, Poignet H, Scatton B (1991) Nitric oxide mediates neuronal death after focal cerebral ischemia in the mouse. Eur J Pharmacol 204:339–340
69. Nagafuji T, Matsui T, Koide T, Asano T (1992) Blockade of nitric oxide formation by N^{ω}-nitro-L-arginine mitigates ischemic brain edema and subsequent cerebral infarction in rats. Neurosci Lett 147:159–162
70. Carreau A, Duval D, Poignet H, Scatton B, Vige X, Nowicki J-P (1994) Neuroprotective efficacy of N^{ω}-nitro-L-arginine after focal cerebral ischemia in the mouse and inhibition of cortical nitric oxide synthase. Eur J Pharmacol 256:241–249
71. Trifiletti RR (1992) Neuroprotective effects of N^G-nitro-L-arginine in focal stroke in the 7-day old rat. Eur J Pharmacol 218:197–198

72. Oury TD, Piantadosi CA, Crapo JD (1993) Cold-induced brain edema in mice: involvement of extracellular superoxide dismutase and nitric oxide. J Biol Chem 268:15394–15398
73. Dawson DA, Kusumoto K, Graham DI, McCulloch J, Macrae IM (1992) Inhibition of nitric oxide synthesis does not reduce infarct volume in a rat model of focal cerebral ischemia. Neurosci Lett 142:151–154
74. Kuluz JW, Prado R, Dietrich WD, Schleien CL, Watson DW (1993) The effect of nitric oxide synthase inhibition on infarct volume after reversible focal cerebral ischemia in conscious rats. Stroke 24:2023–2029
75. Yamamoto S, Golanov EV, Berger SB, Reis DJ (1992) Inhibition of nitric oxide synthesis increases focal ischemic infarction in rat. J Cereb Blood Flow Metab 12:717–726
76. Nagafuji T, Matsui T, Sugiyama M, Koide T, Asano T (1993) Inhibition of nitric oxide synthesis mitigates ischemic brain edema and infarction in rats. J Neurochem 61:S142
77. Nishikawa T, Kirsch J, Koehler RC, Bredt DS, Snyder SH, Traystman R (1993) Effect of nitric oxide synthase inhibition on cerebral blood flow and injury volume during focal ischemia in cats. Stroke 24:1717–1724
78. Richard MS, Knowles RG, Moncada S (1991) Widespread tissue distribution, species distribution and changes in activity of Ca^{2+}-dependent and Ca^{2+}-independent nitric oxide synthases. FEBS Lett 291:145–149
79. Boughton-Smith NK, Evans SM, Laszlo F, Whittle BJR, Moncada S (1993) The induction of nitric oxide synthase and intestinal vascular permeability by endotoxin in the rat. Br J Pharmacol 110:1189–1195
80. Liu S, Adcock IM, Old RW, Baarnes PJ, Evans TW (1993) Lipopolysaccharide treatment in vivo induces widespread tissue expression of inducible nitric oxide synthase mRNA. Biochem Biophys Res Commun 196:1208–1213
81. Liu T, Mc Donnell PC, Young PR, White RF, Siren AL, Hallenbeck JM, Barone FC, Feuerstein GZ (1993) Interleukin-1b mRNA expression in ischemic rat cortex. Stroke 24:1746–1751
82. Frew JD, Paisley K, Martin W (1993) Selective inhibition of basal but not agonist-stimulated activity of nitric oxide in rat aorta by N^G-monomethyl-L-arginine. Br J Pharmacol 110:1003–1008
83. Pou S, Pou WS, Bredt DS, Snyder SH, Rosen GM (1992) Generation of superoxide by purified brain nitric oxide synthase. J Biol Chem 267:24173–24176
84. Dawson DA, Graham DI, McCulloch, Macrae IM (1994) Anti-ischemic efficacy of a nitric oxide synthase inhibitor and a *N*-methyl-D-aspartate receptor antagonist in models of transient and permanent focal cerebral ischemia. Br J Pharmacol 113:247–253
85. Ashwal S, Cole DJ, Osborne TN, Pearce WJ (1993) Low dose L-NAME reduces infarct volume in the rat MCAO/reperfusion model. J Neurosurg Anesthesiol 5:241–249
86. Zhang F, Iadecola C (1993) Nitroprusside improves blood flow and reduces brain damage after focal ischemia. Neuroreport 4:559–562
87. Choi DW (1993) Nitric oxide: foe or friend to the injured brain. Proc Natl Acad Sci USA 90:9471–9743
88. Snyder SH (1993) Janus faces of nitric oxide. Nature 364:577
89. Lei SZ, Pan Z-H, Aggarwal SK, Chen H-SV, Hartman J, Sucher NJ, Lipton SA (1992) Effects of nitric oxide production on the redox modulatory site of the NMDA receptor-channel complex. Neuron 8:1087–1099
90. Lipton SA, Choi Y-B, Pan Z-H, Lei SZ, Chen H-SV, Sucher NJ, Loscalzo J, Singel DJ, Stamler JS (1993) A redox-based mechanism for the neuroprotective and neurodestructive effects of nitric oxide and related nitroso-compounds. Nature 364:626–631
91. Manzoni O, Prezeau L, Marin P, Desharger S, Bockaert J, Fagni L (1992) Nitric oxide-induced blockade of NMDA receptors. Neuron 8:653–662
92. Matsui T, Nagafuji T, Auer RN, Koide T, Tsutsumi K, Asano T, (1993) Beneficial effect of nitric oxide synthase inhibitor on reversible and permanent focal cerebral ischemia in rats, using newly devised miniclip. J Cereb Blood Flow Metab 13:S150

93. Morikawa E, Huang Z, Moskowitz MA (1992) L-Arginine decreases infarct size caused by middle cerebral artery occlusion in SHR. Am J Physiol 263:H1632–H1635
94. Morikawa E, Moskowitz MA, Huang Z, Yoshida T, Irikura K, Dalkara T (1994) L-Arginine infusion promotes nitric oxide-dependent vasodilation, increases regional cerebral blood flow, and reduces infarction volume in the rat. Stroke 25:429–435
95. Zhang F, White JG, Iadecola C (1994) Nitric oxide donors increase blood flow and reduce brain damage in focal ischemia: evidence that nitric oxide is beneficial in the early stage of cerebral ischemia. J Cereb Blood Flow Metab 14:217–226
96. Feelisch M, Noack EA (1987) Correlation between nitric oxide formation during degradation of organic nitrates and activation of guanylate cyclase. Eur J Pharmacol 139:19–30
97. Greenberg SS, Wilcox DE, Rubanyi GM (1990) Endothelium-derived relaxing factor released from canine femoral artery by acethylcholine cannot be identified as free nitric oxide by electron paramagnetic resonance spectroscopy. Circ Res 67:1446–1452
98. Mülsch A, Vanin A, Mordvintcev P, Hauschildt S, Busse R (1992) NO accounts completely for the oxygenated nitrogen species generated by enzymic L-arginine oxygenation. Biochem J 288:597–603
99. Saltzman BE (1954) Colorimetric microdetermination of nitrogen dioxide in the atmosphere. Anal Chem 26:1949–1955
100. Green LC, Wagner DA, Glogowski J, Skipper PL, Wishnok JS, Tannenbaum SR (1982) Analysis of nitrate, nitrite, and [^{15}N]nitrate in biological fluids. Anal Chem 126:131–138
101. Kaspar H, Tiedje JM (1980) Response of electron-capture detector to hydrogen, oxygen, nitrogen, carbon dioxide, nitric oxide and nitrous oxide. J Chromatgr 193:142–147
102. Palmer RMJ, Ferrige AG, Moncada S (1987) Nitric oxide release accounts for the biological activity of endothelium-derived relaxing factor. Nature 327:524–526
103. Pollock JS, Klinghofer V, Förstermann U, Murad F (1992) Endothelial nitric oxide synthase is myristylated. FEBS Lett 309:402–404
104. Komarov A, Mattson D, Jones MM, Singh PK, Lai C-S (1993) In vivo spin trapping of nitric oxide in mice. Biochem Biophys Res Commun 195:1191–1198
105. Akaike T, Yoshida M, Miyamato Y, Sato K, Kohno M, Sasamoto K, Miyazaki K, Ueda S, Maeda H (1993) Antagonistic action of imidazolineoxyl N-oxides against endothelium-derived relaxing factor/·NO through a radical reaction. Biochemistry 32:827–832
106. Shibuki K (1990) An electrochemical microprobe for detecting nitric oxide release in brain tissue. Neurosci Res 9:69–76
107. Shibuki K, Okada D (1991) Endogenous nitric oxide release required for long-term synaptic depression in the cerebellum. Nature 349:326–328
108. Malinski T, Taha Z (1992) Nitric oxide release from a single cell measured in situ by a porphyrinic-based microsensor. Nature 358:676–677
109. Tsukahara H, Gordienko DV, Goligorsky MS (1993) Continuous monitoring of nitric oxide release from human umbilical vein endothelial cells. Biochem Biophys Res Commun 193:722–729
110. Ichimori K, Ishida H, Fukahori M, Nakazawa H, Murakami E (1994) Practical nitric oxide measurement employing a nitric oxide-selective electrode. Rev Sci Instrum 65:2714–2719
111. Matsui T, Nagafuji T, Asano T, Kumura E, Yoshimine T, Hayakawa T (1994) Possible participation of nitric oxide in ischemic brain injury. Adv Neurol Sci (Tokyo) 38:957–966
112. Tominaga T, Sato S, Ohnishi T, Ohnishi T (1993) Potentiation of nitric oxide formation following bilateral carotid occlusion and focal cerebral ischemia in the rat: in vivo detection of the nitric oxide radical by electron paramagnetic resonance spin trapping. Brain Res 614:342–346
113. Kader A, Frazzini VI, Solomon RA, Trifiletti RR (1993) Nitric oxide production during focal cerebral ischemia in rats. Stroke 24:1709–1716

114. Kumura E, Kosaka H, Shiga T, Yoshimine T, Hayakawa T (1994) Elevation of plasma nitric oxide end products during focal cerebral ischemia and reperfusion in the rat. J Cereb Blood Flow Metab 14:487–491
115. Betz AL (1993) Oxygen free radicals and the brain microvasculature. In: Pardridge WM (ed) The blood–brain barrier Raven, New York, pp 303–321
116. Kilbourn RG, Belloni P (1990) Endothelial cell production of nitrogen oxides in response to interferon g in combination with tumor necrosis factor, interleukin-1, or endotoxin. J Natl Cancer Inst 82:772–776
117. Zhang ZG, Chopp M, Gautam S, Zaloga C, Zhang RL, Schmidt HHHW, Pollock JS, Förstermann U (1994) Upregulation of neuronal nitric oxide synthase and mRNA, and selective sparing of nitric oxide synthase-containing neurons after focal cerebral ischemia in rat. Brain Res 654:85–89
118. Zhang ZG, Chopp M, Zaloga C, Garcia JH, Pollock JS, Försterman U (1993) Cerebral endothelial nitric oxide synthase expression after focal cerebral ischemia in rats. Stroke 24:2016–2021
119. Nguyen I, Brunson D, Crespi CL, Penman BW, Wishnok JS, Tannenbaum SR (1992) DNA damage and mutation in human cells exposed to nitric oxide in vitro. Proc Natl Acad Sci USA 89:3030–3034
120. Wink DA, Kasprzak KS, Maragos CM, Elespuru RK, Misra M, Dunams TM, Cebula TA, Koch WH, Andrews AW, Allen JS, Keefer LK (1991) DNA deaminating ability and genotoxicity of nitric oxide and its progenitors. Science 254:1001–1003
121. Malinski T, Bailey F, Zhang G, Chopp M (1993) Nitric oxide measured by a porphyrinic microsensor in rat brain after transient middle cerebral artery occlusion. J Cereb Blood Flow Metab 13:355–358
122. Zhang H, Huang PL, Panahian N, Dalkara T, Fishman MC, Moskowitz MA (1994) Effects of cerebral ischemia in mice deficient in neuronal nitric oxide synthase. Science 265:1883–1885
123. Stamler JS, Singel DJ, Loscalzo J (1992) Biochemistry of nitric oxide and its redox-activated forms. Science 258:1898–1902
124. Zhang J, Snyder SH (1992) Nitric oxide stimulates auto-ADP-ribosylation of glyceraldehyde-3-phosphate dehydrogenase. Proc Natl Acad Sci USA 89:9382–9385
125. Zhang J, Dawson VL, Dawson TM, Snyder SH (1994) Nitric oxide activation of poly(ADP-ribose) synthetase in neurotoxicity. Science 263:687–688
126. Beckman JS (1991) The double-edged role of nitric oxide in brain function and superoxide-mediated injury. J Dev Physiol 15:53–59
127. Saran M, Michel C, Bors W (1990) Reaction of NO with O_2^-. Implications for the action of endothelium-derived relaxing factor. Free Radical Res Commun 10:221–226
128. Radi R, Beckman JS, Bush KM, Freeman BA (1991) Peroxynitrite oxidation of sulfhydryls. J Biol Chem 266:4244–4250
129. Beckman JS, Beckman TW, Chen J, Marshall PA, Freeman BA (1990) Apparent hydroxyl radical production by peroxynitrite: implications for endothelial injury from nitric oxide and superoxide. Proc Natl Acad Sci USA 87:1602–1624
130. Radi R, Beckman JS, Bush KM, Freeman BA (1991) Peroxynitrite-induced membrane lipid peroxidation: the cytotoxic potential of superoxide and nitric oxide. Arch Biochem Biophys 288:481–487
131. Heinzel B, John M, Klatt P, Böhme E, Mayer B (1992) Ca^{2+}/calmodulin-dependent formation of hydrogen peroxide by brain nitric oxide synthase. Biochem J 281:627–630
132. Hogg N, Darley-Usmar VM, Wilson MT, Moncada S (1992) Production of hydroxyl radicals from the simultaneous generation of superoxide and nitric oxide. Biochem J 281:419–424
133. Dumuis A, Sebben GN, Haynes H, Pin J-P, Bockaert J (1988) NMDA receptors activate the arachidonic acid cascade system in striatal neurons. Nature 336:68–70
134. Salvemini D, Misko TP, Masferrer JL, Seibert K, Currie MG, Needleman P (1993) Nitric oxide activates cyclooxygenase enzymes. Proc Natl Acad Sci USA 90:7240–7244

135. Zhu XZ, Luo L-G (1992) Effect of nitroprusside (nitric oxide) on endogenous dopamine release from rat striatal slices. J Neurochem 59:932–935
136. Prast H, Philippu A (1992) Nitric oxide releases acethylcholine in the basal forebrain. Eur J Pharmacol 216:139–140
137. Montague PR, Gancayco CD, Winn MJ, Marchase RB, Friedlande MJ (1994) Role of NO production in NMDA receptor-mediated neurotransmitter release in cerebral cortex. Science 263:973–977
138. Koide T, Asano T, Matushita H, Takakura K (1986) Enhancement of ATPase activity by a lipid peroxide of arachidonic acid in rat brain microvessels. J Neurochem 46:235–242
139. Estrada C, Mengual E, Gonzalez C (1993) Local NADPH-diaphorase neurons innervate pial arteries and lie close or project to intracerebral blood vessels: a possible role for nitric oxide in the regulation of cerebral blood flow. J Cereb Blood Flow Metab 13:978–984
140. Iadecola C, Pelligrio DA, Moskowitz MA, Lassen NA (1994) Nitric oxide synthase inhibition and cerebrovascular regulation. J Cereb Blood Flow Metab 14:175–192
141. Faraci FM, Brian JE (1994) Nitric oxide and the cerebral circulation. Stroke 25:692–703

Protective Effect of Brain Microinjury Against Brain Ischemia

KOKI SHIMOJI, TADAYUKI HONDA, NAOSHI FUJIWARA, KIICHIRO TAGA, SATORU FUKUDA, and SUMIHISA AIDA

Summary. Survival rate following brain ischemia was significantly higher in previously brain-injured mice than in sham-operated mice. Improvement of survival rates in mice with brain injury became insignificant when brain ischemia was imposed under ketamine anesthesia. These results suggest that anti-ischemic factors are released by the injured brain, and also that the anti-ischemic mechanisms developed by brain injury are antagonized by ketamine. On the other hand, the increase in $[Ca^{2+}]_i$ induced by oxygen/glucose deprivation (ischemia in vitro) was inhibited in hippocampal slices from mice with prior (1–3 days) mechanical microinjury at the hippocampal CA1 region. The inhibition of $[Ca^{2+}]_i$ increase was remarkable in the areas close to the injury site. This result also suggests that anti-ischemic factors are produced locally 1–3 days after mechanical brain injury. Thus, certain protective factors against ischemia are suggested to be intrinsically produced in the mechanically injured brain. Elucidation of the intrinsic anti-ischemic activity and its clinical application may lead to new treatments of brain ischemia and brain resuscitation.

Key words. Mechanical brain microinjury—Anti-ischemic activity—Brain ischemia—Neurotrophic factor—Brain slice—Intracellular Ca^{2+} image—Fura-2

Introduction

Although various treatments of brain ischemia to protect the brain against ischemic damage have been attempted, no clinically efficacious treatments have been established. We previously demonstrated that the survival rate following brain ischemia was improved in mice with prior brain injury, suggesting that a certain anti-ischemic activity is induced in the injured brain [1,2]. The phenomenon of neurons pretreated with a short-term (sublethal) ischemia being resistant to ischemia-induced neuronal degeneration, which is called "ischemic tolerance," has also been reported [3,4].

Mechanical injury seems to induce some neurotrophic factor that has been demonstrated in injured brains [5–7]. Although it is unknown whether neurotrophic factors are effective to protect neurons against brain ischemia, some factors induced by mechanical injury may ameliorate the viability of neurons. If some factors are pro-

Department of Anesthesiology, School of Medicine Niigata University, 1-757 Asahimachi-dori, Niigata, 951 Japan.

duced in an injured brain that is resistant to ischemia, they may inhibit the cellular damage induced by ischemia.

An increase in $[Ca^{2+}]_i$ is thought to be the final common path leading to ischemic neuronal damage [8,9]. Thus, measurement of regional $[Ca^{2+}]_i$ in mechanically injured brain tissue might be important for evaluating cellular damage caused by ischemia. In an in vitro study using brain slices, regional changes in $[Ca^{2+}]_i$ of the brain tissue can be determined by a Ca^{2+} image analyzer using fluorescent calcium indicators such as fura-2 [10]. Other physiological parameters can also be recorded under artificial ischemic conditions using brain slices [11,12].

In this chapter, the effect of prior mechanical brain injury on survival following brain ischemia was reviewed; then, regional changes in $[Ca^{2+}]_i$ induced by ischemia were observed using mouse hippocampal slices to characterize the anti-ischemic activity produced in the mechanically injured brain.

Prior Mechanical Brain Injury Improves Survival Following Brain Ischemia

We found previously that mice with a minor injury in the brain were more resistant to brain ischemia than nonoperated and sham-operated mice without brain injury [1,2]. Minor brain injury was inflicted with a small needle at four sites 1 week before the production of incomplete brain ischemia in the mouse. A bilateral carotid clamp was applied for 60 min under pentobarbital anesthesia, and the number of survivors at 1 week after the ischemic insult was compared with that in animals only anesthetized and that in sham-operated mice.

The number of survivors in the brain-injured mice was significantly higher than that in the other two groups of mice. These results suggest that certain unknown protective mechanisms against ischemia become active following mechanical injury of the brain. Improvement of the survival rates in mice with brain injury, however, became insignificant when brain ischemia was imposed during ketamine anesthesia, suggesting that the protective mechanisms against brain ischemia developed by brain injury are antagonized by ketamine or potentiated by barbiturate anesthesia.

Prior Mechanical Brain Microinjury Prevents Ischemia-Induced Elevation of Intracellular Ca^{2+} Concentration

Prior mechanical brain microinjury has been found to have a preventive effect on brain ischemia. If the anti-ischemic activity that increases the survival rate following ischemia is induced in the mechanically injured brain, this anti-ischemic activity may be most efficacious in injured brain tissue. To investigate the mechanism responsible for this, the effect of mechanical brain injury on changes in intracellular free Ca^{2+} concentration ($[Ca^{2+}]_i$) in response to ischemic insult was studied in mouse hippocampal slices.

Mechanical injury was produced by inserting a 25G hypodermic needle into the CA1 region of the hippocampus in mice anesthetized with pentobarbital. Hippocampal slices were then prepared at 2h and at 1, 3, 7, and 14 days after the brain injury. Changes in $[Ca^{2+}]_i$ in the slices were analyzed by fluorescence images using fura-2 [10]. Membrane potentials and membrane resistance of CA1 pyramidal neurons were also measured by an intracellular recording technique.

Extent of Brain Injury

An injury tract was found to pass through the CA1 region and dentate gyrus of the hippocampus (Fig. 1a). The discrepancies between stereotaxic measurement and actual histological examination were minimal in all preparations. Neurons and glial cells disappeared in the injury tract, which was occupied by erythrocytes. However, pyramidal cells were clearly observed in areas at distances of 150 μm and more from the center of the injury tract (Fig. 1b).

Changes in $[Ca^{2+}]_i$ in Response to Ischemia In Vitro

In hippocampal slices from noninjured mice, changes in $R_{340/380}$ of each stratum, which correspond to changes in $[Ca^{2+}]_i$ uniformly increased at 10 min of ischemia in vitro (Fig. 2a). Increase in $R_{340/380}$ of the stratum radiatum (SR) was generally larger than that of the stratum oriens (SO) and stratum pyramidale (SP).

In slices from the injured mice, $R_{340/380}$ was also uniformly increased in each stratum of 0-day slices. Remarkable regional differences were observed in the vicinity of the injury tract 1 and 3 days after the injury (Fig. 2b). The areas nearer to the injury tract showed smaller changes in $R_{340/380}$ in both 1- and 3-day slices. Moreover, the $R_{340/380}$

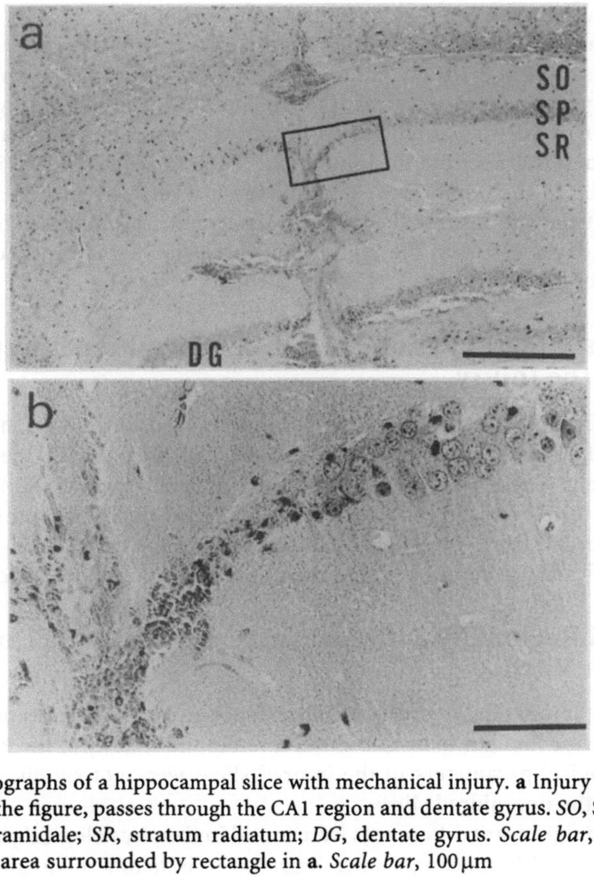

Fig. 1a,b. Micrographs of a hippocampal slice with mechanical injury. **a** Injury tract, observed at the center of the figure, passes through the CA1 region and dentate gyrus. *SO*, Stratum oriens; *SP*, stratum pyramidale; *SR*, stratum radiatum; *DG*, dentate gyrus. *Scale bar*, 500 μm. **b** Expanded view of area surrounded by rectangle in **a**. *Scale bar*, 100 μm

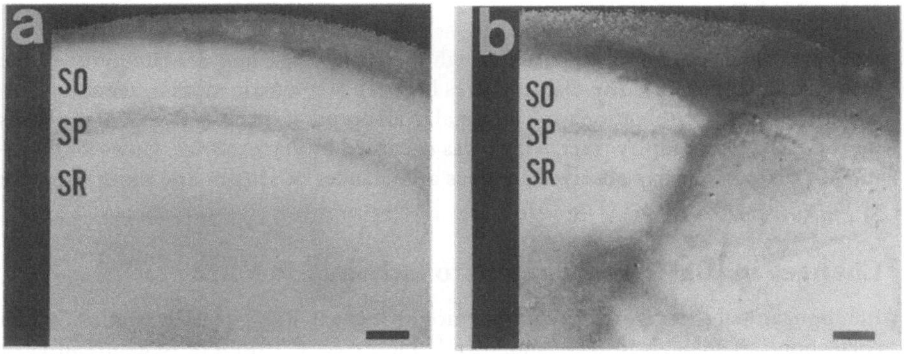

Fig. 2a,b. Ca^{2+} images of the CA1 area of noninjured and injured hippocampal slices; examples of Ca^{2+} image of 3-day slices exposed to 10-min oxygen/glucose deprivation. **a** Noninjured slice; **b** injured slice. Brightness in the images indicates high $[Ca^{2+}]_i$. *Scale bar,* 100 μm

increase in the posterior side across the injury track was further inhibited in comparison with that in the anterior side. The regional differences in $R_{340/380}$ increase became small and insignificant in 7- and 14-day-slices. Differences in the $R_{340/380}$ increase among strata were observed only at 14 days after the injury.

Relative changes in $R_{340/380}$ by 10 min of ischemia in vitro are demonstrated in Fig. 3a. Differences were statistically compared using data recorded from the area surrounded by squares in injured slices (Fig. 3b). As shown in Fig. 3a, significant ($P < 0.05$) inhibition of the $R_{340/380}$ increase was found at every region in 1-day slices and at the area 150 μm posterior from the injury tract (P150) on each stratum in 3-day-slices when compared to the noninjured slices. In these slices, increases in $R_{340/380}$ tend to be inhibited at the site nearest the injury tract. The $R_{340/380}$ increases by ischemia in 0-, 7-, and 14-day slices from the injured mice were not significantly different from those of the noninjured slices.

Electrophysiological Parameters of Cell Membranes in the Vicinity of the Injury

Resting membrane potentials and membrane resistances of CA1 neurons (150–200 μm from the injured tract) were -71 ± 8 mV and 34 ± 12 MΩ (mean ± SD, $n = 13$), respectively. These were not significantly different from those in noninjured slices: -70 ± 6 mV and 44 ± 22 MΩ ($n = 6$), respectively. Waveforms of action potentials recorded from every neuron in response to the depolarizing current were also not significantly different between the injured and noninjured slices.

Analyses of Regionally Different Increases in $[Ca^{2+}]_i$ During Ischemia In Vitro

Effects of brain injury on the resting membrane potential and membrane resistance of CA1 neurons were thought to be minimized even in the vicinity of the injury, because these physiological parameters in the injured slices were not significantly different from those in the noninjured slices.

Although no significant differences were found in the physiological parameters, the increase in $R_{340/380}$ of hippocampal slices exposed to ischemia in vitro for 10 min was significantly inhibited in the vicinity of the injury sites 1 and 3 days after the injury. These regional differences in the $R_{340/380}$ changes did not appear in 0-day slices and were not significant in 7- and 14-day slices from the injured mice. No significant regional difference in $R_{340/380}$ increase was observed in any stratum of the noninjured slice. The time-dependence of the inhibition of $[Ca^{2+}]_i$ increase observed in slices of the injured brain was also significantly different from that in noninjured slices of the noninjured brain. Therefore, it is reasonable to consider that the inhibition of $R_{340/380}$ increase, namely, the inhibition of increase in $[Ca^{2+}]_i$, was produced by injury of the brain tissue and that this inhibitory effect was induced 1–3 days after the injury.

Excessive increases in $[Ca^{2+}]_i$ are thought to be a cause of ischemic brain damage [8,9,13]. In brain slices superfused with a solution deprived of both oxygen and glucose, often used as an in vitro model of brain ischemia, the deleterious effect of calcium accumulation on neuronal function has been demonstrated [14,15]. Indeed, remarkable calcium accumulation was demonstrated in the CA1 area of gerbil hippocampal slices exposed to oxygen/glucose deprivation [16]. Hence, the inhibition of the $[Ca^{2+}]_i$ increase shown in the current experiment suggests that cells in the vicinity of injury were more resistant to ischemia than those in remote regions from the injury tract and those in the slices of the noninjured brain.

The role of glutamate receptors, especially the *N*-methyl-D-aspartate (NMDA) receptor, was emphasized in the $[Ca^{2+}]_i$ increase induced by brain ischemia [8,17,18]. It has been shown that the NMDA receptors are distributed at a high density in the CA1 area of the hippocampus [19], and Schaffer collaterals of CA3 axons have been shown to project to the axial dendrites of CA1 neurons [20]. When fibers are damaged by mechanical injury (as shown in Fig. 1), glutamate release from the nerve terminal may be minimized on the distal side of the injury.

The significantly smaller $[Ca^{2+}]_i$ increase found on the posterior side of the injury (on the distal side of the injury from the CA3) 3 days after the injury suggests that lesions of the presynaptic fibers inhibit, at least in part, the $[Ca^{2+}]_i$ response to oxygen/glucose deprivation. However, the inhibition of $[Ca^{2+}]_i$ increase on the anterior side of the injury 1 and 3 days after injury cannot be explained by injury of the nerve fibers projecting to CA1 neurons. Another, additional mechanism that inhibits $[Ca^{2+}]_i$ increase must be produced in the injured slices during this period.

Improvement of survival rates following brain ischemia has previously been demonstrated in mice 7 days after the brain injury [1], while the $[Ca^{2+}]_i$ increase in 7-day slices of the injured brains was not significantly different from that in the noninjured group. Thus, the mechanism responsible for inhibition of the $[Ca^{2+}]_i$ increase in hippocampal pyramidal cells may not be directly related to improvements in survival following ischemia, as has previously been demonstrated [1]. Some other additional anti-ischemic activity may be promoted in mice 7 days after the brain injury.

Induction of Intrinsic Anti-ischemic activity

Mechanical injury may induce changes of biosynthesis in the injured region. Development of various neurotrophic factors has been demonstrated. For example, the induction of neurotrophic factors has been demonstrated in injured brains [5–7,21]. Enhancement in maintenance of cellular Ca^{2+} homeostasis has also been demonstrated as one of their important functions [6]. If such mechanisms are endogeneously

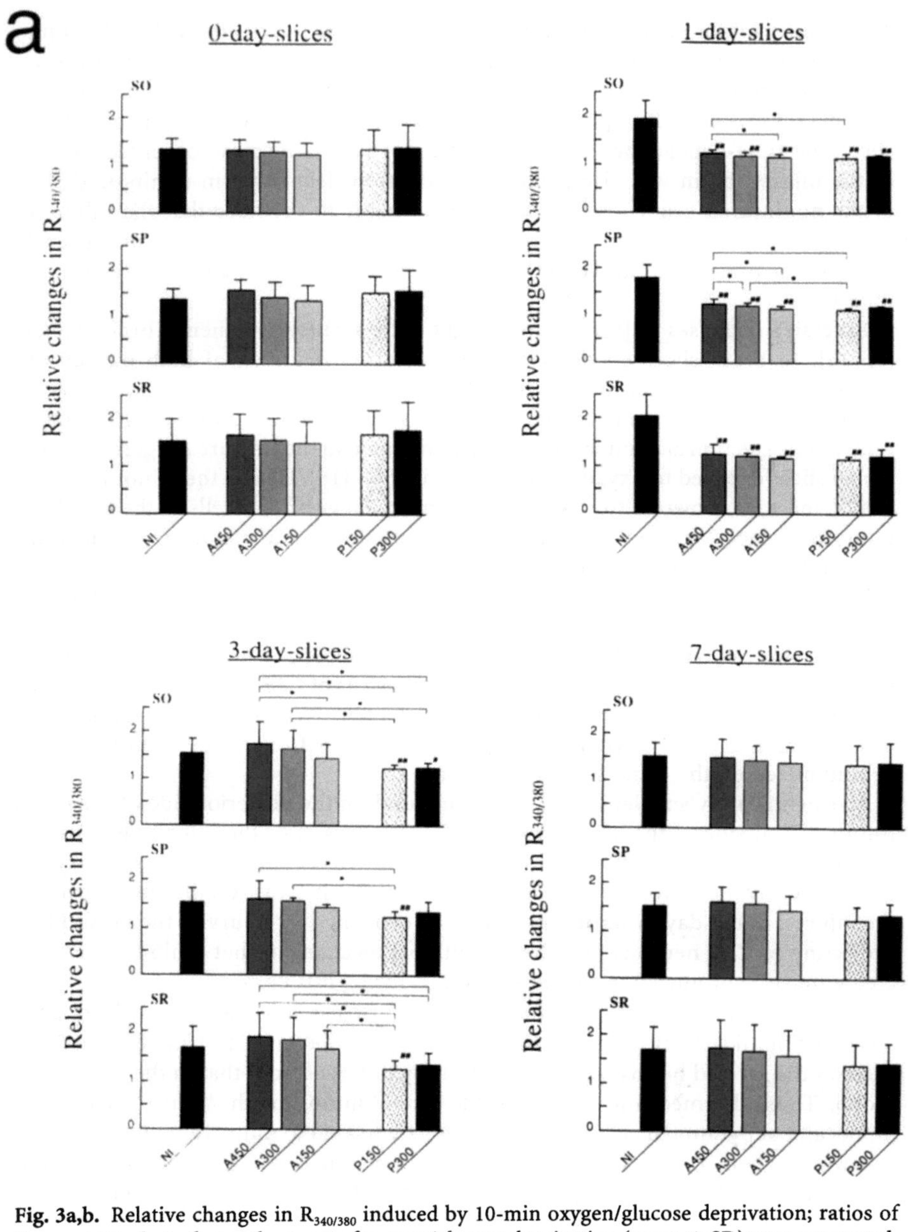

Fig. 3a,b. Relative changes in $R_{340/380}$ induced by 10-min oxygen/glucose deprivation; ratios of $R_{340/380}$ at 10 min to that at the onset of oxygen/glucose deprivation (mean ± SD) were compared. **a** *NI*, Non-injured group; *A150*, *A300*, *A450*, *P450*, and *P300* indicate the positions for $R_{340/380}$ measurements as shown in **b**, where 40 µm × 40 µm square windows are located on each stratum at 150-, 300-, and 450-µm-separated positions on the anterior side (A150, A300, A450) and at 150- and 300-µm-separated positions on the posterior side (P150, P300) from the injury tract. *Scale bar*, 100 µm. *, $P < .05$ between positions in the injured slice; #, $P < .05$; ##, $P < .01$, in comparison with NI

14-day-slices

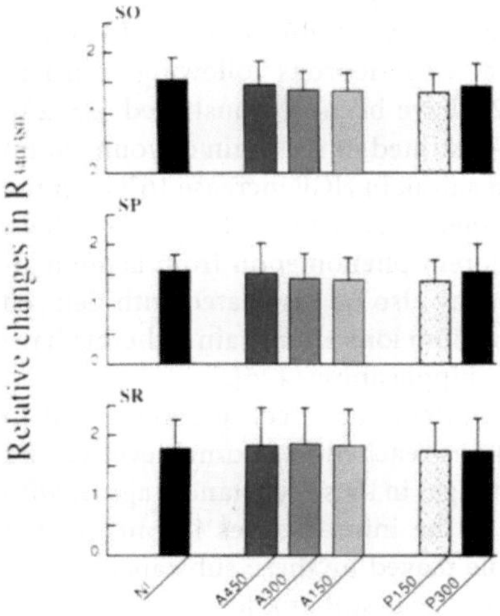

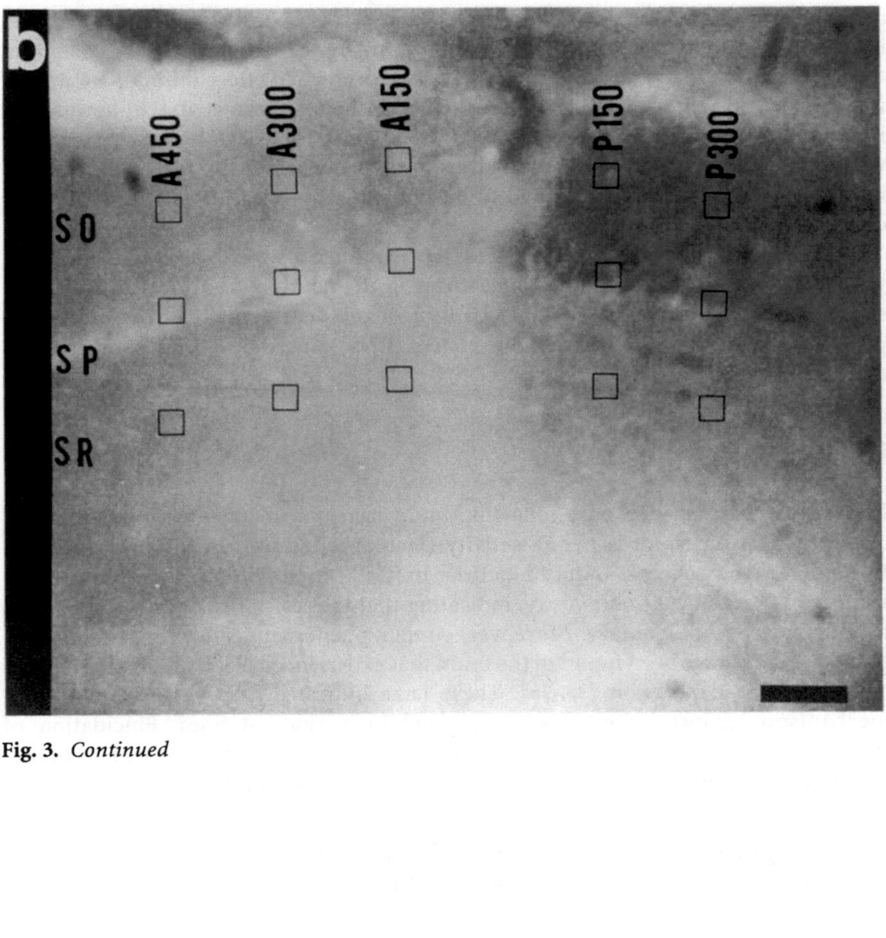

Fig. 3. *Continued*

produced by traumatic brain injury, their effects could also be efficacious in the injured brain itself, especially in the vicinity of the injury [1,2].

Protective effects of exogeneously administered nerve growth factor (NGF) at high doses against the degeneration of hippocampal CA1 neurons following transient ischemia, so-called delayed neuronal death [22], have been demonstrated [23–25]. Time-dependent increase in NGF has been demonstrated in the brain of young adult rats after cortical injury [5]. These authors found a peak in NGF increase 16h after the injury; a relatively high level of NGF was sustained during the following 3–6 days. Although delayed neuronal death may be a different phenomenon from acute neuronal damage, excessive calcium accumulation may also be associated with delayed neuronal death. Indeed, sustained increases in calcium ions after brain ischemia have been demonstrated in the CA1 area of the gerbil hippocampus [26].

Induction of heat shock proteins and *c-fos* proteins has been demonstrated in injured brains [27]; the expression of *c-fos* proteins reached a maximal level 12 and 24h after the trauma. The time-dependence of changes in these substances agrees with that of the inhibition of the $[Ca^{2+}]_i$ response in the injured slices in our current experiment. Unfortunately, the physiological role played by these substances in the intracellular calcium changes induced by brain injury is still unknown.

Stress to the brain as well as mechanical injury may also cause anti-ischemic activity. Several studies have demonstrated that brief ischemia (for 2min), which in itself has no lethal effect on neurons, induces tolerance to subsequent ischemia in the hippocampal neurons in gerbils [3,4]. The protective effect induced by a brief ischemia lasted for at least 4 days after the first ischemia. Taga et al. [28] have found in rats that induced cortical spreading depression (CSD) leads to an adaptive response which reduces the vulnerability of neurons to subsequent ischemic injury. This CSD-induced protective effect was found to be evident 24h after CSD and lasted for 3–7 days.

These reports suggest that minor stress to the brain causes it to produce changes in gene expression and biochemical substances in the neurons to protect itself from a subsequent ischemic insult. The mechanism of the development of these intrinsic protective activities in central nervous system (CNS) caused by a brief insult remains to be determined.

Conclusion

Improvement of survival rate following brain ischemia in mice with brain injury suggests that some anti-ischemic activity (factor) is induced in the injured brain. Inhibition of the ischemia-induced increase in $[Ca^{2+}]_i$ in brain slices was shown in the vicinity of mechanical brain injury, indicating that the resistance of cells to ischemia is enhanced by brain injury. Moreover, similar phenomena showing resistance to brain ischemia have been found in the brain that experienced prior sublethal ischemia and spreading depression [3,4,28]. These facts indicated that intrinsic protective mechanisms against ischemia were induced by various stresses. Elucidation of the intrinsic antiischemic activity and its clinical application may provide a breakthrough to treatments of brain ischemia and brain resuscitation.

Acknowledgment. This work was supported in part by a grant-in-aid from the Japanese Ministry of Education, Science and Culture.

References

1. Takahata Y, Shimoji K (1986) Brain injury improves survival of mice following brain ischemia. Brain Res 381:368–371
2. Shimoji K, Takahata Y, Fujiwara N, Endoh H, Taga K, Ohama E (1987) Effects of pentobarbital and ketamine on brain injury-induced anti-ischemic activity. Brain Res 408:385–388
3. Kato H, Liu Y, Araki T, Kogure K (1991) Temporal profile of the effects of pretreatment with brief cerebral ischemia on the neuronal damage following secondary ischemic insult in the gerbil: cumulative damage and protective effects. Brain Res 553:238–242
4. Kirino T, Tsujita Y, Tamura A (1991) Induced tolerance to ischemia in gerbil hippocampal neurons. J Cereb Blood Flow Metab 11:299–307
5. Ishikawa R, Nishikori K, Furukawa S (1991) Appearance of nerve growth factor and acidic fibroblast growth factor with different time courses in the cavity-lesioned cortex of the rat brain. Neurosci Lett 127:70–72
6. Mattson MP, Scheff SW (1994) Endogeneous neuroprotection factors and traumatic brain injury: mechanism of action and implications for therapy. J Neurotrauma 11:3–33
7. Nieto-Sampedro M, Lewis ER, Cotman CW, Manthorpe P, Skaper SD, Barbin G, Longo FM, Varon S (1982) Brain injury causes a time-dependent increase in neurotrophic activity at the lesion site. Science 217:860–861
8. Choi DW (1988) Calcium-mediated neurotoxicity: relationship to specific channel types and role in ischemic damage. Trends Neurosci 11:465–469
9. Deshpande JK, Siesjö BK, Wieloch T (1987) Calcium accumulation and neuronal damage in the rat hippocampus following cerebral ischemia. J Cereb Blood Flow Metab 7:89–95
10. Kudo Y, Ito K, Miyakawa H, Izumi Y, Ogura A, Kato H (1987) Cytoplasmic calcium elevation in hippocampal granule cell induced by perforant path stimulation and L-glutamate application. Brain Res 407:168–172
11. Dong WQ, Schurr A, Reid KH, Shields CB, West CA (1988) The rat hippocampal slice preparation as an in vitro model of ischemia. Stroke 19:498–502
12. Fujiwara N, Higashi H, Shimoji K, Yoshimura M (1987) Effects of hypoxia on rat hippocampal neurones *in vitro*. J Physiol (Camb) 387:131–151
13. Siesjö BK (1981) Cell damage in the brain: a speculative synthesis. J Cereb Blood Flow Metab 1:155–185
14. Lipton P, Lobner D (1990) Mechanisms of intracellular calcium accumulation in the CA1 region of rat hippocampus during anoxia in vitro. Stroke 21:III60–III64
15. Schurr A, Rigor BM (1989) Cerebral ischemia revisited: new insight as revealed using in vitro brain slice preparations. Experientia (Basel) 45:684–695
16. Mitani A, Kadoya F, Kataoka K (1990) Distribution of hypoxia-induced calcium accumulation in gerbil hippocampal slice. Neurosci Lett 120:42–45
17. Benveniste H, Jørgensen MB, Diemer NH, Hansen AJ (1988) Calcium accumulation by glutamate receptor activation is involved in hippocampal cell damage after ischemia. Acta Neurol Scand 78:529–536
18. Choi DW (1990) The role of glutamate neurotoxicity in hypoxic-ischemic neuronal death. Annu Rev Neurosci 13:171–182
19. Monagham DT, Holets VR, Toy DW, Cotman CW (1983) Anatomical distributions of four pharmacologically distinct ^{3}H-L-glutamate binding sites. Nature 306:176–179
20. Lopes da Silva FH, Arnolds DEAT (1978) Physiology of the hippocampus and related structures. Annu Rev Physiol 40:185–216
21. Manthorpe M, Nieto-Sampedro M, Skaper SD, Lewis ER, Barbin G, Longo FM, Cotman CW, Varon S (1983) Neurotrophic activity in brain wounds of the developing rat. Correlation with implant survival in the cavity. Brain Res 267:47–56
22. Kirino T (1982) Delayed neuronal death in the gerbil hippocampus following ischemia. Brain Res 239:57–69

23. Buchan AM, Williams L, Bruederlin B (1990) Nerve growth factor: pretreatment ameliorates ischemic hippocampal neuronal injury. Stroke 21:177
24. Shigeno T, Mima T, Takakura K, Graham DI, Kato G, Hashimoto Y, Furukawa S (1991) Amelioration of delayed neuronal death in the hippocampus by nerve growth factor. J Neurosci 11:2914–2919
25. Yamamoto S, Yoshimine T, Fujita T, Kuroda R, Irie T, Fujioka K, Hayakawa T (1992) Protective effect of NGF atelocollagen minipellet on the hippocampal delayed neuronal death in gerbils. Neurosci Lett 141:161–165.
26. Hashimoto K, Kikuchi H, Ishikawa M, Kobayashi S (1992) Changes in cerebral energy metabolism and calcium level in relation to delayed neuronal death after ischemia. Neurosci Lett 137:165–168
27. Dragunow M, Robertson HA (1988) Brain injury induces c-fos protein(s) in nerve and glial-like cells in adult mammalian brain. Brain Res 455:295–299
28. Taga K, Patel JC, Drummond JC, Cole DJ (1995) Pre-ischemic cortical spreading depression reduces cortical injury in rats subjected to forebrain ischemia. J Cereb Blood Flow Metab 15:S65

Subject Index

MIX
Papier aus verantwortungsvollen Quellen
Paper from responsible sources
FSC® C105338

If you have any concerns about our products,
you can contact us on
ProductSafety@springernature.com

In case Publisher is established outside the EU,
the EU authorized representative is:
Springer Nature Customer Service Center GmbH
Europaplatz 3, 69115 Heidelberg, Germany

Printed by Libri Plureos GmbH
in Hamburg, Germany